# Disclaimer

The publisher of this book is by no way associated with the National Institute of Standards and Technology (NIST). The NIST did not publish this book. It was published by 50 page publications under the public domain license.

50 Page Publications.

**Book Title:** Feature Information Model for Disassembly

**Book Author:** Dale P. Bentz; Chiara F. Ferraris; Kenneth A. Snyder;

**Book Abstract:** As sustainability becomes an important issue nowadays, disassembly becomes one of the key processes to turn out-of-service products into new or rebuilt products to save energy and materials. A disassembly process includes many subprocesses, such as separation, cleaning, and inspection. Features play a key role in design for disassembly and disassembly process planning. This paper describes a NIST-developed feature information model that defines generic form features, pattern features, and disassembly-specific features. The information model provides a foundation for computer-aided design for disassembly software systems development and data exchange among heterogeneous systems to integrate product design and disassembly process planning.

**Keywords:** Disassembly Feature; Feature Model; and Information Modeling

# Feature Information Model for Disassembly

Shaw C. Feng[1], Thomas Kramer[1], Che B. Joung[1], Parisa Ghodous[2], and Milton Borsato[1,3]

[1]National Institute of Standards and Technology (NIST)
[2]University of Claude Bernard Lyon I
[3]Federal University of Technology – Parana

## Abstract

As sustainability becomes an important issue nowadays, disassembly becomes one of the key processes to turn out-of-service products into new or rebuilt products to save energy and materials. A disassembly process includes many subprocesses, such as separation, cleaning, and inspection. Features play a key role in design for disassembly and disassembly process planning. This paper describes a NIST-developed feature information model that defines generic form features, pattern features, and disassembly-specific features. The information model provides a foundation for computer-aided design for disassembly software systems development and data exchange among heterogeneous systems to integrate product design and disassembly process planning.

**Key words:**
Disassembly Feature, Feature Model, and Information Modeling.

## 1. Introduction

Disassembly is a process to separate functional parts from an out-of-service product. The National Institute of Standards and Technology (NIST) has developed an information model of features to support product disassembly. An open disassembly feature model will enable designers to specify disassembly features in the design stage. The model can also help process planners use features in determining disassembly methods and sequences. This information model supports both the structure of a disassembled product and its disassembly process. A disassembly process includes separating parts from a product, inspecting the quality of the separated parts for reuse or remanufacturing, and cleaning reusable parts. Disassembly planning requires information from design, such as disassembly features and their relationships, subassemblies, and disassembly sequence. Disassembly process planning systems need also to exchange disassembly process plans with other design and process planning systems. Designers require the cost of disassembly and available equipment to determine the disassemblability of a design. An information model is, hence, developed to meet the needs of exchanging disassembly data between design and disassembly process planning.

This paper describes a disassembly feature information model, developed using the Unified Modeling Language (UML) [1]. Section 2 reviews disassembly feature-related literature. Section 3 describes the classes and their relationships that are comprised of the information model. Section 4 concludes the paper and provides possible future direction.

## 2. Review of Disassembly Feature-related Literature

This section provides a literature review of disassembly representation in design and feature modeling. Gaps in an integrated disassembly information model are identified at the end of the section.

**2.1** Design for disassembly

Many research results have been published in the areas of design for disassembly, disassembly process planning, and cost estimation. Vinodh, Praveen Kumar, and Nachiappan [2] present methods for modeling planning and leveling the disassembly process of a cam-operated rotary switch assembly. The methodology involves disassembly modeling using a graphical approach, based on the work done by Tang, Zhou and Caudill [3].

Mascle and Zhao [4] describe a general methodology in Design for Environment (DFE) using entropy minimization. The entropy evaluation brings about the generation of a disassembly sequence in which the disassembly efficiency, the material value and the specific value are big and the liability is small, as a gold ship's chronometer. Fuzzy logic and feature modeling are used during the DFE evaluation for parts, assembly and operations analysis.

Behdad and Thurston [5] address the problem of disassembly sequence planning for the purposes of maintenance or component upgrading. Optimization of the disassembly sequence is carried out while simultaneously considering tradeoffs among attributes: disassembly time (and resulting cost), the probability of no part damage and the reversibility of disassembly sequence (both time and the probability of not incurring damage during reassembly).

Fan, et al. [6] evaluate the recycling rates, costs as well as the disassembly time of a notebook at its end-of-life stage using data collected during disassembly processes.

Liu, et al. [7] apply an improved max–min ant system-based algorithm to the problem of product disassembly sequence planning.

Smith, Smith and Chen [8] present a new 'disassembly sequence structure graph' (DSSG) model for multiple-target selective disassembly sequence planning. The DSSG model contains a minimum set of parts, with an order and direction for removing each part. The approach uses expert rules to choose parts, part order, and part disassembly directions, based upon physical constraints.

Tseng, Chang and Cheng [9] propose a disassembly-oriented assessment method for product modular design in stages, focusing upon the green design issue in terms of the disassembly and recycle of product modules.

Li, Wang and Huang [10] suggest an integrated approach of disassembly constraint generation, based on which an object-oriented prototype is designed and developed. It makes uses of CAD standard component library and geometric constraint information for precedence and non-precedence (geometric) constraint generation.

2

## 2.2 Disassembly-related information models

This section presents a review of available assembly information models for the purpose of creating a disassembly information model. Some standard-based approaches and frameworks are also reviewed.

ISO 10303-44 [11] provides some limited assembly design representations that capture assembly structure and kinematic joint information during the design process. The assembly model establishes a neutral representation of product assembly. In this model, a product is called an "assembly," and the components of the lowest level in an assembly are called "parts." The model focuses on the hierarchy of a product and on the position and orientation between the parts.

The Open Assembly Model (OAM) [12] has the aim to provide a standard representation and exchange protocol for assembly and system-level assembly information. OAM is extensible. It currently provides feature representation and propagation, representation of kinematics, and engineering analysis at the system level. The assembly information model emphasizes the information requirements for part features and assembly relationships. The model includes both assembly as a concept and assembly as a data structure. For the latter, it uses model data structures of STEP.

Vinodh, Nachiappan and Praveen Kumar [13] apply four different modeling strategies for efficiently representing all feasible and complete disassembly sequences with correct precedence relations: connection graph, directed graph, AND/OR graph and disassembly Petri net (DPN). A case example of a rotary switch is used and the advantages and drawbacks of each approach are discussed. The connection graph portrays the local and the global constraints among the components, but is incapable of modeling disassembly tasks. The directed graph represents sets of all disassembly sequences, but as the number of components increases the size of this also increases significantly. The AND/OR graph is also a directed graph which includes reduced number of nodes and edges. It also shows the simultaneous execution of disassembly tasks. However, it is difficult to integrate with resource modeling. The DPN represents each node and the task involved between them individually. This graph requires the component-fastener, which makes it accurate, but complex.

Lambert and Gupta [14] present the state diagram, which is structured with nodes and arcs, where the nodes are the different states in the disassembly sequence, and the arcs are the operations to separate the components. Each state has free components and components that will still be connected. The first state brings the whole equipment or product with all components connected and the last state presents all free components, ready to use or for final disposal.

## 2.3 Feature information for disassembly

Most feature representations are related to assembly and machining. An assembly feature can be defined as a shape feature of a part in an association with a shape feature of another part in an assembly. Many research results on assembly feature modeling are applicable to disassembly.

Shah et al. [15, 16] describe the association between faying features of parts. Their work deals with the determination of geometric constraints: degrees of freedom, compatibility between faying features, orientation, and insertion limits. Chan et al. [17] define an assembly feature as the elementary connection feature containing faying relations between the components. Hamidullah et al. [18] define assembly features and their representation using the concept of assembly intents. They specify the information of assembly, i.e., faying relations with the connecting form features, and associate the connecting form features with other assembly-specific information, such as assembly operations and assembly degrees of freedom.

ISO 10303-224 [19] defines product data necessary for manufacturing a single piece or assembly of mechanical parts. It has machining features such as hole and pocket, and transition features such as round, fillet, and chamfer. ISO 10303-111 [20] otherwise specifies resource constructs for representing the complex shape elements, also known as form features, that are supported by the solid modeling capabilities of modern CAD systems. ISO 10303-111 defines depression features such as hole and pocket, and edge blended features, such as edge blend and chamfered edge.

Dipper, Xu and Klemm [21] take a feature data model, i.e., ISO 10303-238 [22], as input and determine interactions among the features in the model. The feature interaction data are then appended to the original data model for subsequent uses such as process planning.

Hu, et al. [23] propose an approach for manufacturability evaluation based on feature modeling using an object-oriented methodology. A gear box is used as a case to illustrate the approach towards virtual manufacturing and concurrent engineering.

Anjum, et al. [24] present a shape feature-based ontological model of the geometry of engineering components for analyzing their manufacturability in early design stages. A case study demonstrates how such a technique can be used to overcome the issue of semantic inconsistency.

**2.4** Summary of high-level requirements for an information model for disassembly

Based on the above literature review, the following gaps are identified for modeling the information of integrated design for disassembly and disassembly process planning:
- Comprehensive modeling on simple, compound, and pattern disassembly features and separation between features from two component parts.
- Modeling complex relationships among various features in disassembly.

## 3. Disassembly Feature Information Model

This section describes the classes and their relationships in the disassembly feature information model in UML class diagrams. The model has two major packages (i.e., modules). The Support Data Package will be described in Section 3.1, and the Feature Package will be described in Section 3.2.

4

Throughout this model, the object being disassembled is called a workpiece, and it is assumed that the workpiece has a coordinate system.

**3.1** Support Data

The Support Data Package consists of a set of data classes and one subpackage PlacementPack. The purpose of the Support Data Package is to support classes in other packages in the disassembly information model. Default data types, such as character (**char**), **boolean**, **integer**, **double**, and **float**, in the UML are used throughout the model. Figure 1 shows the diagram of all the classes and the PlacementPack sub-package of the Support Data Package.

(Figure 1 goes here.)

3.1.1 Support types

Class String is used to represent a list of one or more characters. It has one attribute. Attribute chars is a list of ***char***[1] of the type of Character, which is defined in the UML.

Class Identification is used to represent the identification of an object. It has one attribute. Attribute *theID*[2] is of type String.

Class ExplicitItem[3] is used to represent the identification of a piece of geometry on a solid model, such as a cylindrical surface representing a hole that is identified as a feature. The class has one attribute. Attribute *itemID* represents the identification of a piece of geometry. The attribute's data type is String.

Class Measure is an abstract data type. It has subclasses of LengthMeasure, MeasureWithUnit, and AngularMeasure.

Enumeration type LengthUnitType includes five commonly used length units, *km, m, cm, mm,* and *micron.*

Class GlobalLengthUnit is used to represent the global length unit. It has one attribute. Attribute *unit* represents the global unit of length measure. Its data type is LengthUnitType.

Class LengthMeasure is used to represent the measurement of a length. It has one attribute. Attribute *value* represents the value of the length measure, and its data type is ***double***, which is a UML data type. The length unit is specified in an instance of GlobalLengthUnit.

Enumeration type AngularUnitType includes two commonly used angular units, *radian* and *degree.*

---

[1] A type in bold italic font denotes a UML defined type.
[2] An attribute of a class in the disassembly information model is in italic font.
[3] The first letter of each word in a class name is capitalized.

Class GlobalAngularUnit is used to represent the global angular unit in the disassembly model. It has one attribute. Attribute *unit* represents the global unit of angular measure.

Class AngularMeasure is used to represent the measurement of an angle. It has one attribute. Attribute *value* represents the value of angular measure, and its data type is **double**, which is a UML data type. The angular unit is specified in an instance of GlobalAngularUnit.

Class MeasureWithUnit is used to represent the measurement of a general measurand. It has two attributes. Attribute *unit* represents the unit of measure, and its data type is String. Attribute *value* represents the value of the measure, and its data type is **double**, which is a UML data type.

Class PositiveInteger is used to represent an integer that is greater than zero. It has one attribute. Attribute *x* represents the positive integer, and its data type is **int**, which is a UML data type. There is a constraint that *x* should be greater than zero.

Class Coordinates3D is used to represent coordinates of a point in the 3D space. It has three attributes. Attributes *x*, *y*, and *z* represent the 3 coordinates in the space, and their data types are **double**.

Class UnitVector3D is used to represent a unit vector in the 3D space. It has three attributes. Attributes *x*, *y*, and *z* represent the 3 components of the unit vector, and their data types are **double**. There is a constraint that the magnitude of the vector should be one.

Class Vector3D is used to represent a vector in the 3D space. It has two attributes. Attribute *origin* represents the starting point of the vector, and its data type is Coordinates3D. Attribute *end* represents the end point of the vector, and its data type is also Coordinates3D. The *origin* and *end* are not coincident.

Class PointAndDirection is used to represent a point with a direction that is used in an inspection process. It has two attributes. Attribute *point* represents the point, and its data type is Coordinates3D. Attribute *direction* represents the direction, and its data type is UnitVector3D.

Class Plane is used to represent a plane in the 3D space. It has two attributes. Attribute *location* represents the location point of the plane, and its data type is Coordinates3D. Attribute *normalVector* represents the orientation of the plane, and its data type is UnitVector3D.

3.1.2 Placements

In this model, stereotypical geometric entities (points, curves, and surfaces) representing the shape of a workpiece in 3D space are built by defining each type of entity in a native 3D Cartesian coordinate system and then placing that coordinate system in the coordinate system of a workpiece using a 3DPlacementZX, a 3DPlacementZ, or a 3DPlacement. The planar construction geometry is placed on the XY plane of a 3D coordinate system, and the method of using the construction geometry is described. For each geometric entity, a mathematical description of the set of points on the entity in its native coordinate system is given. Figure 2 shows a diagram of all the classes of placement.

(Figure 2 goes here.)

Class 3DPlacement is used to represent the placement of a 3D coordinate system in a 3D space. It has one attribute. Attribute *location* represents the location of the origin of the coordinate system in the space, and its data type is Coordinates3D. The directions of the axes of the placed coordinate system are not specified in this class but are in its subclasses.

Class 3DPlacementZ is a subclass of 3DPlacement. It is used to represent the location and partial orientation of a 3D coordinate system in a 3D space. In addition to the *location* inherited from 3DPlacement, it has the attribute *zDirection,* which represents the direction of the Z axis of the coordinate system. Its data type is Vector3D. The direction of the X axis is not specified in this class but is in its subclass.

Class 3DPlacementZX is a subclass of 3DPlacementZ. It is used to represent the location and complete orientation of a 3D coordinate system in a 3D space. In addition to the *location* and *zDirection* inherited from 3DPlacementZ, it has the attribute *xDirection,* which represents the direction of the X axis of the coordinate system. Its data type is Vector3D. The *xDirection* vector must be perpendicular to the *zDirection* vector. The Y axis direction of the coordinate system is determined by the right-hand rule as the cross product of the *zDirection* vector and the *xDirection* vector.

## 3.2 Feature

A disassembly feature represents a specified portion of the surface of a workpiece that is of interest in disassembly, including separation, cleaning, and inspection. A feature can be a point, a planar feature, or a fully three-dimensional feature. Figure 3 shows a diagram of all the high-level classes and subpackages in the disassembly feature model.

(Figure 3 goes here.)

Class DisassemblyFeature is a subclass of the OAM:Feature class since it contains basic attributes that are applicable to the disassembly feature. Class OAM:Feature is used to represent a feature defined in the Open Assembly Model [12]. DisassemblyFeature has two attributes. Attribute *ID* represents the feature identification, and its data type is Identification. Attribute *explicitRepresentation* represents the geometric and topological items in a boundary representation of the workpiece that coincide with the DisassemblyFeature, and its data type is ExplicitItem. ExplicitItem is defined in ISO 10303-203 [25].The *explicitRepresentation* of a DisassemblyFeature includes exactly those edges and surfaces of the workpiece that might be seen or felt. The feature descriptions given in this class are more idealized than the *explicitRepresentation* since they ignore any intersecting features that might remove portions of a curve or surface. In general, to determine if any given point is actually present on a feature, it is necessary to use the *explicitRepresention.*

Many features may be either inside or outside of material. These features have an *insideXxx* attribute of optional **boolean** type. A value of true means the feature is inside of material (the

7

surface of a hole, for example). A value of false means the feature is outside of material (the surface of peg, for example).

3.2.1 Feature Classes

This section introduces four abstract subclasses of DisassemblyFeature: DirectPlacementFeature, PatternElementFeature, IndirectPlacementFeature, and ImplicitPlacementFeature. They are defined according to the method used to locate a feature on a workpiece. The PatternElementFeature class exists to allow efficient representations of PatternFeatures while making it straightforward to treat any feature in a pattern as an independent feature.

Class DirectPlacementFeature is an abstract subclass of DisassemblyFeature. It has one additional attribute. Attribute *placement* , which is of type 3DPlacement (or one of its subtypes), locates the DirectPlacementFeature on the workpiece.

Class PatternElementFeature is an abstract subclass of DisassemblyFeature. It has three additional attributes. Attribute *patternFeatureID*, which is of type Identification, gives the identifier of the PatternFeature of which the PatternElementFeature is a member. Attribute *firstIndex*, which is of type **integer**, gives the first index of an element of the PatternFeatureShape in the PatternFeature. Attribute *secondIndex* gives the second index of the same element of the PatternFeatureShape in the PatternFeature. Each subclass of PatternFeatureShape includes a description of how elements of the pattern are indexed. For MatrixPatternFeatureShape and its subclasses *firstIndex* is a row index and *secondIndex* is a column index. For CircularPatternFeatureShape and its subclasses, *firstIndex* is the index around the circular arc and the value of *secondIndex* (which is still required) is not used. For ConcentricCircularPatternFeatureShape and its subclasses, *firstIndex* is the index around an arc and *secondIndex* is the index of the arc.

Class IndirectPlacementFeature is an abstract subclass of DisassemblyFeature. The placement of an IndirectPlacementFeature is determined by the placement(s) of one or more DirectPlacementFeature(s) associated with the IndirectPlacementFeature. Many form features (e.g., HoleFeature) are IndirectPlacementFeatures.

Class ImplicitPlacementFeature is an abstract subclass of DisassemblyFeature. An ImplicitPlacementFeature modifies one or more other features. The placement of an ImplicitPlacementFeature is determined by the placement(s) of the modified feature(s) and the nature of the modification. Chamfer, Fillet, RoundEdge, and Thread features are subclasses of ImplicitPlacementFeature.

3.2.2 FeatureShapes and DirectPlacementFeatures

As shown in Figure 4, Class FeatureShape describes the shape of a feature. It is abstract. The subclasses of FeatureShape cover common geometric shapes. Each FeatureShape has a native Cartesian coordinate system. That system requires no explicit representation. The descriptions of subclasses of FeatureShape in this paper describe where the FeatureShape is in its native coordinate system and how the parameters of the FeatureShape are used to determine its shape.

Some FeatureShapes, such as PointFeatureShape and SphereFeatureShape (Figure 9), do not require being oriented, but may be oriented. Instances of FeatureShapes are placed using a 3DPlacement, 3DPlacementZ, or 3DPlacementZX. Other FeatureShapes, such as LineFeatureShape, CircleFeatureShape (Figure 6), and PlaneFeatureShape (Figure 7), must be partially oriented but are not required to be completely oriented. These instances of FeatureShapes are placed using a 3DPlacementZ or 3DPlacementZX. Most features are required to be completely oriented. They are placed using a 3DPlacementZX.

(Figure 4 goes here.)

The following section consists of pairs of paragraphs. The first paragraph of the pair describes a FeatureShape, and the second paragraph of the pair describes a DirectPlacementFeature that uses the FeatureShape. The section is divided further by the dimensionality of the lowest dimensional subspace into which a feature may be placed: Point Shape and Feature (0D), Linear Shapes and Features (1D), Planar Shapes and Features (2D), and 3D Shapes and Features.

## 3.2.2.1 Point Shape and Feature

Class PointFeatureShape is a subclass of FeatureShape used to represent a point that is used in a disassembly process, such as inspection of a feature on a workpiece. In its native coordinate system, a PointFeatureShape is the single point at the origin, (0, 0, 0).

Class PointFeature is a subclass of DirectPlacementFeature used to represent a point positioned in a 3D space. PointFeature has one additional attribute. Attribute *pointShape* represents the point and its type is PointFeatureShape. The *placement* must be a 3DPlacement, a 3DPlacementZ, or a 3DPlacementZX.

## 3.2.2.2 Linear Shapes and Features

Class LineFeatureShape is a subclass of FeatureShape used to represent a line bounded at one end. It has no attributes. The line lies on the Z axis of its native coordinate system starting at the origin and extends indefinitely along the positive Z axis. The points of a LineFeatureShape are all points represented by (0, 0, z) for $0 <= z$.

Class LineFeature is a subclass of DirectPlacementFeature used to represent a line positioned in a 3D space. LineFeature has one additional attribute. Attribute *LineShape* represents the line shape, and its type is LineFeatureShape. The *placement* must be a 3DPlacementZ or a 3DPlacementZX.

Class BoundedLineFeatureShape is a subclass of FeatureShape used to represent a line segment bounded at both ends. It has one attribute. Attribute *length* represents the length of the line segment, and its type is LengthMeasure. The line lies on the Z axis of its native coordinate system starting at the origin and extends to (0, 0, length). The points of a LineFeatureShape are all points represented by (0, 0, z) for $0 <= z <= length$.

Class BoundedLineFeature is a subclass of DirectPlacementFeature used to represent a line segment positioned in a 3D space. BoundedLineFeature has one additional attribute. Attribute *boundedLineShape* represents the bounded line shape, and its type is BoundedLineFeatureShape. The *placement* must be a 3DPlacementZ or a 3DPlacementZX.

3.2.2.3 Planar Shapes and Features

Figure 5 shows the diagram of all the classes of planar linear features. As mentioned earlier, planar features lie on the XY plane of a 3D coordinate system.

(Figure 5 goes here.)

Class PlanarCurveFeatureShape is an abstract subclass of FeatureShape that is a parent class for several subclasses of planar shapes that are curves.

Class PlanarCurveDirectPlacementFeature is an abstract subclass of DirectPlacementFeature that is a parent class for several subclasses of planar features whose shape is a PlanarCurveFeatureShape.

Class ParallelLinesFeatureShape is a subclass of PlanarCurveFeatureShape used to represent a pair of parallel line segments on the XY plane. It has three attributes. Attribute *length* represents the length of the line segments, and its data type is LengthMeasure. Attribute *distance* represents the distance between the two parallel lines, and its data type is LengthMeasure. Attribute *insideParallelLines* represents whether the parallel line segments are inside of material or outside of material. Its data type is optional **boolean**. A value of true indicates that the lines are inside of material, e.g., lines on the sides of a slot. A value of false indicates that the lines are outside of material, e.g., lines on the sides of a rib. In the native coordinate system of a ParallelLinesFeatureShape, one line segment lies on the X axis starting at the origin (0, 0, 0) and ending at (*length*, 0, 0), and the other line segment lies on the XY plane starting at (0, *distance*, 0) and ending at (0, *distance*, *length*). The points of a Parallel3DLinesFeatureShape are all points represented by (x, 0, 0) or (x, *distance*, 0) for $0 <= x <= length$.

Class ParallelLinesFeature is a subclass of PlanarCurveDirectPlacementFeature used to represent a pair of parallel line segments positioned in a 3D space. ParallelLinesFeature has one additional attribute. Attribute *parallelLinesShape* represents the shape of the parallel line segments, and its type is ParallelLinesFeatureShape. The *placement* must be a 3DPlacementZX.

Class ParallelLinesFlatEndFeatureShape is a subclass of PlanarCurveFeatureShape used to represent the shape that results when the ends of the parallel lines in a ParallelLinesFeatureShape with the same X coordinates are connected to each other by two parallel line segments. The shape is the outline of a rectangle. ParallelLinesFlatEndFeatureShape has all the attributes of a ParallelLinesFeatureShape, and there are no additional attributes. The points of a ParallelLinesFlatEndFeatureShape are all the points of a ParallelLinesFeatureShape plus all points represented by (0, y , 0) or (*length*, y, 0) for $0 <= y <= distance$.

Class ParallelLinesFlatEndFeature is a subclass of PlanarCurveDirectPlacementFeature used to represent the outline of a rectangle positioned in a 3D space. ParallelLinesFlatEndFeature has one additional attribute. Attribute *parallelLinesFlatEndShape* represents the shape of the rectangle, and its type is ParallelLinesFlatEndFeatureShape. The *placement* must be a 3DPlacementZX.

Class ParallelLinesRoundEndFeatureShape is a subclass of PlanarCurveFeatureShape used to represent the shape that results when the ends of the parallel lines in a ParallelLinesFeatureShape with the same X coordinates are connected to each other by semicircles. Each round end is a semicircle whose diameter equals the distance between the two parallel lines. The semicircles make the feature convex, not concave. ParallelLinesRoundEndFeatureShape has all the attributes of a ParallelLinesFeatureShape, and there are no additional attributes. The points of a ParallelLinesRoundEndFeatureShape are all the points of a ParallelLinesFeatureShape plus:
1. all points represented by ($length$ + R*cos(a), R+ R*sin(a), 0) for $-\pi/2 < a < \pi/2$ and
2. all points represented by (R*cos(a), R+ R*sin(a), 0) for $\pi/2 < a < 3\pi/2$

where R = *distance*/2.

Class ParallelLinesRoundEndFeature is a subclass of PlanarCurveDirectPlacementFeature used to represent a pair of parallel lines joined by semicircles at the ends positioned in a 3D space. ParallelLinesRoundEndFeature has one additional attribute. Attribute *parallelLinesRoundEndShape* represents the shape of the feature, and its type is ParallelLinesRoundEndFeatureShape. The *placement* must be a 3DPlacementZX.

Figure 6 shows the diagram of all the classes of circular features.

(Figure 6 goes here.)

Class CircleFeatureShape is a subclass of PlanarCurveFeatureShape used to represent a curve that is a circle in the XY plane. It has two attributes. Attribute *radius* represents the radius of the circle, and its data type is LengthMeasure. Attribute *insideCircle* represents whether the circle is inside of material or outside of material. Its data type is optional **boolean**. True represents that the circle is inside of material, e.g., a circle on the inside of a hole. In the native coordinate system of the circle, the center of the circle is at the origin (0, 0, 0). The points of a CircleFeatureShape are all points represented by (*radius*\*cos(a), *radius*\*sin(a), 0) for $0 <= a < 2\pi$.

Class CircleFeature is a subclass of PlanarCurveDirectPlacementFeature used to represent a circle positioned in a 3D space. CircleFeature has one additional attribute. Attribute *circleShape* represents the circle shape, and its type is CircleFeatureShape. The *placement* must be a 3DPlacementZ or a 3DPlacementZX.

Class ArcFeatureShape is a subclass of PlanarCurveFeatureShape used to represent a circular arc on the XY plane. It has three attributes. Attribute *angle* represents the angle of the arc, and its data type is AngularMeasure. The *angle* must be positive and less than $2\pi$. Attribute *radius* represents the radius of the circle, and its data type is LengthMeasure. Attribute *insideArc* represents whether the arc is inside of material or outside of material. Its data type is optional *boolean*. True represents that the arc is inside of material, e.g., an arc on the inside of a hole. In the native coordinate system of the arc, the center of the arc is at the origin (0, 0, 0). Viewed from the positive Z axis, the arc starts at (radius, 0, 0) and proceeds counterclockwise through the given *angle*. The points of an ArcFeatureShape are all points represented by (*radius*\*cos(a), *radius*\*sin(a), 0) for 0 <= a <= *angle*.

Class ArcFeature is a subclass of PlanarCurveDirectPlacementFeature used to represent an arc positioned in a 3D space. ArcFeature has one additional attribute. Attribute *arcShape* represents the arc shape, and its type is ArcFeatureShape. The *placement* must be a 3DPlacementZX.

Class EllipseFeatureShape is a subclass of PlanarCurveFeatureShape used to represent a curve that is an ellipse on the XY plane. It has three attributes. Attribute *majorDiameter* represents the major diameter of the ellipse, and its data type is LengthMeasure. Attribute *minorDiameter* represents the minor diameter of the ellipse, and its data type is LengthMeasure. Attribute *insideEllipse* represents whether the ellipse is inside of material or outside of material. Its data type is optional *boolean*. True represents that the ellipse is inside of material, e.g., an ellipse on the inside of a hole. In the native coordinate system of the ellipse, the center of the ellipse is at the origin (0, 0, 0) , the *majorDiameter* lies on the X axis, and the *minorDiameter* lies of the Y axis. The points of an EllipseFeatureShape are all points represented by ($R$\*cos(a), $r$\*sin(a), 0) for 0 <= a < $2\pi$, where $R$ = *majorDiameter*/2, and r = *minorDiameter*/2.

Class EllipseFeature is a subclass of PlanarCurveDirectPlacementFeature used to represent an ellipse positioned in a 3D space. EllipseFeature has one additional attribute. Attribute *ellipseShape* represents the ellipse shape, and its type is EllipseFeatureShape. The *placement* must be a 3DPlacementZX.

Figure 7 shows the diagram of all the classes of general curve features and a plane feature.

(Figure 7 goes here.)

Class PlanarGCurveFeatureShape is a subclass of PlanarCurveFeatureShape used to represent a general curve feature in the XY plane. It has one attribute. Attribute *pointsOnPlane* represents the control points that are used to generate the curve, and its data type is a list of Coordinates3D with minimum of four points in the list. The Z coordinate of all control points must be 0. A uniform cubic B-spline curve is assumed.

Class PlanarGCurveFeature is a subclass of PlanarCurveDirectPlacementFeature used to represent a general planar curve positioned in a 3D space. PlanarGCurveFeature has one additional attribute. Attribute *planarGCurveShape* represents the shape of the general curve, and its type is PlanarGCurveFeatureShape. The *placement* must be a 3DPlacementZX.

Class ClosedPlanarGCurveFeatureShape is a subclass of PlanarGCurveFeatureShape used to represent a closed-end general curve feature on the XY plane. No particular order of continuity is required at the point of closure.

Class ClosedPlanarGCurveFeature is a subclass of PlanarCurveDirectPlacementFeature used to represent a closed general planar curve positioned in a 3D space. ClosedPlanarGCurveFeature has one additional attribute. Attribute *closedPlanarGCurveShape* represents the shape of the closed general curve, and its type is ClosedPlanarGCurveFeatureShape. The *placement* must be a 3DPlacementZX.

Class PolyPlanarCurveFeatureShape is a subclass of PlanarCurveFeatureShape used to represent an open or closed continuous curve in the XY plane composed of segments each of which is a planar curve. There must be at least one segment. A PolyPlanarCurveFeatureShape has two attributes. Attribute *segments* represents the segments of the curve, and its data type is list of PlaneCurveFeatureShape. Each segment must be a BoundedLineFeatureShape or a PlanarCurveFeatureShape, excluding ParallelLinesFeatureShape and PolyPlanarCurveFeatureShape itself. Attribute *placements* represents the placements of the individual segments in the native coordinate system of the PolyPlanarCurveFeatureShape, and its data type is list of 3DPlacementZX. The two lists must have the same length. The nth placement gives the placement of the nth segment. The *placements* must be such that the end point of each segment (except the last) is the start point of the next segment. The *placements* must also be such that all segments lie in the XY plane. To ensure that, it is necessary and sufficient that (1) the Z value of the *location* of each placement must be 0, (2) the Z value of the *zDirection* of the placement for a BoundedLineFeatureShape must be 0, and (3) the *zDirection* of the placement for a PlanarCurveFeatureShape must be (0, 0, 1). If any segment is a closed feature (a circle or an ellipse, for example), there must be only one segment. The PolyPlanarCurveFeatureShape must not intersect itself.

Class PolyPlanarCurveFeature is a subclass of PlanarCurveDirectPlacementFeature used to represent an open or closed continuous planar curve composed of segments positioned in a 3D space. PolyPlanarCurveFeature has one additional attribute. Attribute *polyPlanarCurveShape* represents the shape of the curve, and its type is PolyPlanarCurveFeatureShape. The *placement* must be a 3DPlacementZX.

Class PlaneFeatureShape is a subclass of FeatureShape used to represent the surface of a rectangle shape. It has two attributes. Attribute *length* represents the length of the plane, and its data type is LengthMeasure. Attribute *width* represents the width of the plane, and its data type is also LengthMeasure. In its native coordinate system, the rectangle lies on the XY plane. The center of the rectangle is at the origin, length is parallel to the X axis, and width is parallel to the Y axis. The points of the plane are all points represented by (x, y, 0) for
-length/2 <= x <= length/2 and -width/2 <= y <= width/2.

Class PlaneFeature is a subclass of DirectPlacementFeature used to represent a feature that is a plane positioned in a 3D space. PlaneFeature has one additional attribute. Attribute *planeShape* represents the shape of the plane, and its type is PlaneFeatureShape. The *placement* must be a 3DPlacementZ or a 3DPlacementZX.

3.2.2.4 3D Shapes and Features

Figure 8 shows the diagram of 3D general curve features.

(Figure 8 goes here.)

Class GCurve3DFeatureShape is a subclass of FeatureShape used to represent the shape of a feature that is a general 3D curve. It has one attribute. Attribute *pointsOfTheCurve* represents the control points that are used to generate the curve, and its data type is a list of Coordinates3D with minimum of four elements in the list. A uniform cubic B-spline curve is assumed.

Class GCurve3DFeature is a subclass of DirectPlacementFeature used to represent a feature that is a general 3D curve positioned in a 3D space. GCurve3DFeature has one additional attribute. Attribute *gCurve3DShape* represents the shape of the curve, and its type is GCurve3DFeatureShape. The *placement* must be a 3DPlacementZX.

Class ClosedGCurve3DFeatureShape is a subclass of GCurveFeature3DShape used to represent the shape of a closed-end general 3D curve. No particular order of continuity at the closing end is assumed.

Class ClosedGCurve3DFeature is a subclass of DirectPlacementFeature used to represent a feature that is a closed general 3D curve positioned in a 3D space. ClosedGCurve3DFeature has one additional attribute. Attribute *closedGCurve3DShape* represents the shape of the curve, and its type is ClosedGCurve3DFeatureShape. The *placement* must be a 3DPlacementZX.

Class 3DSurfaceFeatureShape is an abstract subclass of FeatureShape used to represent the shape (but not the placement) of a simple 3D feature that is a surface.
Figure 9 shows the diagram of point symmetric features.

(Figure 9 goes here.)

Class PointSymmetricFeatureShape is an abstract subclass of 3DSurfaceFeatureShape used to represent a feature shape that is a surface symmetric about a point placed in a 3D space. The point is the center of the feature and is located at the origin of the native coordinate system of the PointSymmetricFeatureShape. A PointSymmetricFeatureShape has the property that whenever a point (x, y, z) is a point of the feature shape, so is (-x, -y, -z).

Class SphereFeatureShape is a subclass of PointSymmetricFeatureShape used to represent a feature that is a sphere. SphereFeatureShape has two attributes. Attribute *diameter* represents the diameter of the sphere, and its data type is LengthMeasure. Attribute *insideSphere* represents whether the sphere is inside of material or outside of material, and its data type is optional **boolean**. True represents that the sphere is inside of material, i.e., the surface of a spherical void. In its native coordinate system, the center of the sphere is at the origin. The points of a SphereFeatureShape are all points represented by (x, y, z) with $x^2 + y^2 + z^2 = R^2$, where R = *diameter*/2. The representation given for SphereSegmentFeatureShape will also work for the points of a SphereFeatureShape.

Class SphereFeature is a subclass of DirectPlacementFeature used to represent a sphere positioned in a 3D space. SphereFeature has one additional attribute. Attribute *sphereShape*

14

represents the shape of the sphere, and its type is SphereFeatureShape. The *placement* must be a 3DPlacement, a 3DPlacementZ, or a 3DPlacementZX.

Class SphereSegmentFeatureShape is a subclass of SphereFeatureShape used to represent a feature shape that is a spherical segment bounded by lines of longitude and lines (or points) of latitude, where lines of latitude and longitude are arranged as on the earth. The longitude direction is 0 on the positive X axis and increases rotating towards the positive Y axis. The latitude direction is 0 on the XY plane and is positive rotating towards the positive Z axis. The class has all the attributes of a SphereFeatureShape plus three additional attributes. Attribute *startInLatitude* represents the angle of the start of the spherical segment in the latitude direction, and its data type is AngularMeasure. Attribute *endInLatitude* represents the angle of the end of the spherical segment in the latitude direction, and its data type is AngularMeasure. The angles must satisfy
$-\pi/2 <= startInLatitude < endInLatitude <= \pi/2$. If the *startInLatitude* is $-\pi/2$, the segment is bounded by the south pole (0, 0, -R), where R is the radius; otherwise, it is bounded by a line of latitude. If the *endInLatitude* is $\pi/2$, the segment is bounded by the north pole (0, 0, R); otherwise, it is bounded by a line of latitude. The spherical segment starts in longitude at 0. Attribute *endInLongitude* represents the angle of the end of the spherical segment in the longitude direction, and its data type is AngularMeasure. The *endInLongitude* must satisfy $0 < endInLongitude < 2\pi$. The points on the spherical segment are all points represented by (R*cos(a)*cos(b), R*cos(a)*sin(b), R*sin(a)), where a and b are angles, for *startInLatitude* <= a <= *endInLatitude* and 0 <= b <= *endInLongitude*.

Class SphereSegmentFeature is a subclass of DirectPlacementFeature used to represent a sphere segment positioned in a 3D space. SphereSegmentFeature has one additional attribute. Attribute *sphereSegmentShape* represents the shape of the sphere segment, and its type is SphereSegmentFeatureShape. The *placement* must be a 3DPlacementZX.

Class EllipsoidFeatureShape is a subclass of PointSymmetricFeatureShape used to represent a feature shape that is an ellipsoid. EllipsoidFeature has four attributes. Attribute *diameterInX* represents the diameter of the ellipsoid in the X direction, and its data type is LengthMeasure. Attribute *diameterInY* represents the diameter of the ellipsoid in the Y direction, and its data type is LengthMeasure. Attribute *diameterInZ* represents the diameter of the ellipsoid in the Z direction, and its data type is LengthMeasure. Attribute *insideEllipsoid* represents whether the ellipsoid is inside of material or outside of material, and its data type is optional **boolean**. True represents that the ellipsoid is inside of material, i.e., an ellipsoidal void. The points of an EllipsoidFeatureShape are all points represented by (x, y, z) such that
$(x^2/a^2 + y^2/b^2 + z^2/c^2) = 1$, where a = *diameterInX*/2, b = *diameterInY*/2, and c = *diameterInZ*/2.

Class EllipsoidFeature is a subclass of DirectPlacementFeature used to represent an ellipsoid positioned in a 3D space. EllipsoidFeature has one additional attribute. Attribute *ellipsoidShape* represents the shape of the ellipsoid, and its type is EllipsoidFeatureShape. The *placement* must be a 3DPlacementZX.

Figure 10 shows the diagram of axial-symmetric features.

(Figure 10 goes here.)

Class AxisSymmetricFeatureShape is an abstract subclass of 3DSurfaceFeatureShape used to represent a feature shape symmetric about the Z axis. It has no attributes. An AxisSymmetricFeatureShape has the property that the shape looks the same when the feature is rotated around the Z axis by any amount. In mathematical terms whenever a point (x, y, z) is a point of the feature shape, so is (R*cos(a), R*sin(a), z), where R = sqrt($x^2 + y^2$) and a may have any value.

Class AxialFeature is an abstract subclass of DirectPlacementFeature used to represent a feature that references an AxisSymmetricFeatureShape.

Class ConeFeatureShape is a subclass of AxisSymmetricFeatureShape used to represent a feature that is a cone. It has three attributes. Attribute *base* represents the diameter of the cone base, and its data type is LengthMeasure. Attribute *coneAngle* represents the full angle of the cone (not the half angle), and its data type is AngularMeasure. The *coneAngle* must be positive and less than $\pi$ radians. Attribute *insideCone* represents whether the cone is inside of material or outside of material, and its data type is **boolean**. If the value is false, the cone is the surface of a protrusion, otherwise, the cone is the surface of a depression. In its native coordinate system, the axis of the cone is the Z axis, the cone has its base on the XY plane, and the tip of the cone is on the positive Z axis. The points of a ConeFeatureShape are all points represented by
((R – z*tan(A))*cos(b), ((R – z*tan(A))*sin(b), z) where R = *base*/2, A = *coneAngle*/2,
$0 <= z <= R/\tan(A)$, and $0 <= b < 2\pi$.

Class ConeFeature is a subclass of AxialFeature used to represent a cone positioned in a 3D space. ConeFeature has one additional attribute. Attribute *coneShape* represents the shape of the cone, and its type is ConeFeatureShape. The *placement* must be a 3DPlacementZ or a 3DPlacementZX.

Class ConeSegmentFeatureShape is a subclass of ConeFeatureShape used to represent a feature shape that is a cone segment. The cone segment is bounded on the sides by lines from the tip of the cone to the base. The beginning line of the cone segment is the line from the tip of the cone to the point where the circle bounding the base crosses the positive X axis of the native coordinate system of the cone segment. ConeSegmentFeatureShape has all the attributes of a ConeFeatureShape plus one additional attribute. Attribute *endAngleOfSegment* represents the angle of the cone segment, and its data type is AngularMeasure. The *endAngleOfSegment* is the angle on the XY plane between the X axis and the line from the origin to the point where the second bounding line intersects the base. The angle is positive in the counterclockwise direction as viewed from the positive Z axis. The points of a ConeSegmentFeatureShape are all points represented by
((R – z*tan(A))*cos(b), ((R – z*tan(A))*sin(b), z) where R = *base*/2, A = *coneAngle*/2,
$0 <= z <= R/\tan(A)$, and $0 <= b <= $ *endAngleOfSegment* $< 2\pi$.

Class ConeSegmentFeature is a subclass of DirectPlacementFeature used to represent a cone segment positioned in a 3D space. It has one additional attribute. Attribute *coneSegmentShape*

16

represents the shape of the cone segment, and its type is ConeSegmentFeatureShape. The *placement* must be a 3DPlacementZX.

Class FrustumFeatureShape is a subclass of ConeFeatureShape used to represent a feature that is the side of a frustum. FrustumFeatureShape has one additional attribute. Attribute *top* represents the diameter of the top of the frustum, and its data type is LengthMeasure. Its value must be less than *base*. The points of a FrustumFeatureShape are all points represented by $((R - z*\tan(A))*\cos(b), ((R - z*\tan(A))*\sin(b), z)$ where R = *base*/2, A = *coneAngle*/2, $0 <= b < 2\pi$, and $0 <= z <= (base - top)/(2*\tan(A))$.

Class FrustumFeature is a subclass of AxialFeature used to represent a frustum positioned in a 3D space. FrustumFeature has one additional attribute. Attribute *frustumShape* represents the shape of the frustum, and its type is FrustumFeatureShape. The *placement* must be a 3DPlacementZ or a 3DPlacementZX.

Class FrustumSegmentFeatureShape is a subclass of FrustumFeatureShape used to represent a feature shape that is a frustum segment. The frustum segment is bounded on the sides by line segments from the upper circle of the frustum to the base ciricle. The frustum segment is bounded above by an arc of the circle that bounds the top of the frustum. The frustum segment is bounded below by an arc of the circle that bounds the base of the frustum. The beginning line segment bounding the frustum segment is the line to the point where the circle bounding the base crosses the positive X axis of the native coordinate system of the frustum segment. FrustumSegmentFeatureShape has all the attributes of a FrustumFeatureShape plus one additional attribute. Attribute *endAngleOfSegment* represents the angle of the frustum segment, and its data type is AngularMeasure. The *endAngleOfSegment* is the angle on the XY plane between the X axis and the line from the origin to the point where the second bounding line intersects the base. The angle is positive in the counterclockwise direction as viewed from the positive Z axis. The points of a FrustumSegmentFeatureShape are all points represented by $((R - z*\tan(A))*\cos(b), ((R - z*\tan(A))*\sin(b), z)$ where R = *base*/2, A = *coneAngle*/2, $0 <= z <= (base - top)/(2*\tan(A))$, and $0 <= b <= endAngleOfSegment <= 2\pi$.

Class FrustumSegmentFeature is a subclass of DirectPlacementFeature used to represent a frustum segment positioned in a 3D space. FrustumSegmentFeature has one additional attribute. Attribute *frustumSegmentShape* represents the shape of the frustum segment, and its type is FrustumSegmentFeatureShape. The *placement* must be a 3DPlacementZX.

Figure 11 shows the diagram of the other axial-symmetric features.

(Figure 11 goes here.)

Class CylinderFeatureShape is a subclass of AxisSymmetricFeatureShape used to represent a feature shape that is a cylinder. It has three attributes. Attribute *diameter* represents the diameter of the cylinder, and its data type is LengthMeasure. Attribute *insideCylinder* represents whether the cylinder is inside of material or outside of material, and its data type is optional **boolean**. True represents that the cylinder is inside of material, e.g., the surface of a cylindrical hole. Attribute *length* represents the length of the cylinder, and its data type is LengthMeasure. In its native coordinate system, the cylinder is bounded below by the XY plane and above by the plane z = *length*. The points of a CylinderFeatureShape are all points represented by (R*cos(a), R*sin(a), z), where R = *diameter*/2, $0 <= a < 2\pi$, and $0 <= z <= length$.

Class CylinderFeature is a subclass of AxialFeature used to represent a cylinder positioned in a 3D space. CylinderFeature has one additional attribute. Attribute *cylinderShape* represents the shape of the cylinder, and its type is CylinderFeatureShape. The *placement* must be a 3DPlacementZ or a 3DPlacementZX.

Class CylinderSegmentFeatureShape is a subclass of CylinderFeatureShape used to represent a feature shape that is a cylindrical segment. The segment extends the full height of the cylinder, but it is bounded by two line segments on the side of the cylinder. The first line segment is the line parallel to the Z axis passing through the point where the positive X axis passes through the cylinder. The second line segment is also parallel to the Z axis; it hits the XY plane on the circle that bounds the bottom of the cylinder. The CylinderSegmentFeatureShape class has all the attributes of a CylinderFeatureShape plus one additional attribute. Attribute *angleOfSegment* represents the angular size of the cylinder segment, and its data type is AngularMeasure. The *angleOfSegment* is the angle on the XY plane between the X axis and the line from the origin to the point where the second bounding line intersects the XY plane. The angle is positive in the counterclockwise direction as viewed from the positive Z axis. The points of a CylinderSegmentFeatureShape are all points represented by (R*cos(a), R*sin(a), z) where R = *diameter*/2, $0 <= a <= angleOfSegment < 2\pi$, and $0 <= z <= length$.

Class CylinderSegmentFeature is a subclass of DirectPlacementFeature used to represent a cylinder segment positioned in a 3D space. CylinderSegmentFeature has one additional attribute. Attribute *cylinderSegmentShape* represents the shape of the cylinder segment, and its type is CylinderSegmentFeatureShape. The *placement* must be a 3DPlacementZX.

Class TorusFeatureShape is a subclass of AxisSymmetricFeatureShape used to represent a feature shape that is a torus. In its native coordinate system, a torus is generated by rotating a circle in the ZX plane with its center on the X axis (call it the small circle) around the Z axis. The center of the small circle sweeps through a circle on the XY plane (call it the large circle). The TorusFeatureShape class has three attributes. Attribute *majorRadius* represents the radius of the large circle, and its data type is LengthMeasure. Attribute *minorRadius* represents the radius of the small circle, and its data type is LengthMeasure. Attribute *insideTorus* represents whether the torus is inside of material or outside of material, and its data type is optional **boolean**. True represents that the torus is inside of material, i.e., a toroidal void. The *majorRadius* must be larger than the *minorRadius*. The points of the torus are all points represented by ((R + r*cos(a))*cos(b), (R + r*cos(a))*sin(b), r*sin(a)) for $0 <= a <= 2\pi$ and $0 <= b <= 2\pi$, where R = *majorRadius* and r = *minorRadius*.

Class TorusFeature is a subclass of AxialFeature used to represent a torus positioned in a 3D space. TorusFeature has one additional attribute. Attribute *torusShape* represents the shape of the torus, and its type is TorusFeatureShape. The *placement* must be a 3DPlacementZ or a 3DPlacementZX.

Class TorusSegmentFeatureShape is a subclass of TorusFeatureShape used to represent a feature shape that is a torus segment. TorusSegmentFeatureShape has all the attributes of a TorusFeatureShape plus three additional attributes, all of which are of type AngularMeasure, namely *endInMajorDirection*, *startInMinorDirection*, and *endInMinorDirection*. The segment is bounded on four sides by arcs of circles. One pair of opposite sides is arcs of circles, one of which is the original small circle, and the other of which is a rotated copy of the small circle. In the native coordinate system of the torus, the first small circle is on the ZX plane with its center at (R, 0, 0). The second small circle is rotated about the Z axis through an angle *endInMajorDirection* from the first circle. That angle is the angle between the line from the origin of the native coordinate system to the center of the first small circle and the line from the origin to the center of the second small circle. The angle is positive counterclockwise as viewed from the positive Z axis. The direction of that angle may be called the longitude direction. The other pair of opposite sides of the segment is arcs of circles in planes that are parallel to the XY plane. The positions of each of those sides may be described by the angle around the first small circle of the point at which the small circle and the bounding circle intersect. The direction of that angle may be called the latitude direction. Latitudes on a torus differ from latitudes on a sphere in that on a torus they go through a full circle ($2\pi$ radians), not half a circle ($\pi$ radians). The 0 latitude is set at the point farthest from the Z axis where the first small circle intersects the XY plane. The positive latitude direction is towards the positive Z axis from the 0 latitude. The *startInMinorDirection* is the latitude at which the segment starts; it must be positive and less than $2\pi$. The *endInMinorDirection* is the latitude at which the segment ends; it must be greater than the *startInMinorDirection* by some positive amount d. The amount d must be less than $2\pi$. Note that the value of *endInMinorDirection* may be almost $4\pi$. The points of the TorusSegmentFeatureShape are all points represented by
$((R + r*\cos(a))*\cos(b), (R + r*\cos(a))*\sin(b), r*\sin(a))$ for
$0 <= startInMinorDirection <= a <= endInMinorDirection$ and
$0 <= b <= endInMajorDirection < 2\pi$, subject to (*startInMinorDirection* $< 2\pi$) and
$((endInMinorDirection - startInMinorDirection) < 2\pi)$.

Class TorusSegmentFeature is a subclass of DirectPlacementFeature used to represent a torus segment positioned in a 3D space. TorusSegmentFeature has one additional attribute. Attribute *torusSegmentShape* represents the shape of the torus segment, and its type is TorusSegmentFeatureShape. The *placement* must be a 3DPlacementZX.

Class SurfaceOfRevolutionFeatureShape is a subclass of AxisSymmetricFeatureShape used to represent a surface that is created by revolving a planar curve about an axis through a full circle. It has two attributes. Attribute *profile* represents the planar curve, and its data type is PolyPlanarCurveFeatureShape. Attribute *insideRevolutionShape* represents whether the surface of revolution is inside of material or outside of material, and its data type is optional **boolean**. True represents that the surface of revolution is inside of material, e.g., the surface of an irregular hole. In its native coordinate system, no point of the *profile* may have a negative X value, and only the first and last points may have x = 0. To form the surface of revolution, the X axis of the native coordinate system of the *profile* is placed coincident with the X axis of the native coordinate system of the SurfaceOfRevolutionFeatureShape, and the Y axis of the native coordinate system of the *profile* is placed coincident with the Z axis of the native coordinate

19

system of the SurfaceOfRevolutionFeatureShape. Then the *profile* (which is now in the ZX plane of the native coordinate system of the SurfaceOfRevolutionFeatureShape) is rotated in a full circle about the Z axis of the native coordinate system of the SurfaceOfRevolutionFeatureShape. The surface consists of all points through which the *profile* sweeps. Letting the *profile* be represented parametrically by z = 0, x = f(a), and y = g(a) for 0 <= a <=1, those are all points represented by

(f(a)*cos(b), f(a)*sin(b), g(a)) for $0 <= b < 2\pi$.

Class SurfaceOfRevolutionFeature is a subclass of AxialFeature used to represent a surface of revolution positioned in a 3D space. SurfaceOfRevolutionFeature has one additional attribute. Attribute *surfaceOfRevolutionShape* represents the shape of the surface of revolution, and its type is SurfaceOfRevolutionFeatureShape. The *placement* must be a 3DPlacementZ or a 3DPlacementZX.

Figure 12 shows the diagram of cuboid and an elliptical cylinder features.

(Figure 12 goes here.)

Class CuboidFeatureShape is a subclass of 3DSurfaceFeatureShape used to represent a box shape missing the top and bottom. CuboidFeatureShape has four attributes. The first three are *length*, *width*, and *height,* and the data type of those three is LengthMeasure. Attribute *insideCuboid* represents whether the box is inside of material or outside of material, and its data type is optional **boolean**. True represents that the box is inside of material, i.e., a hole with a rectangular cross section. In its native coordinate system, the box has its edges parallel to the coordinate axes, sits on the XY plane extending toward the positive Z axis, and has the middle of the bottom on the origin. Attribute *length* represents the size of the edges of the box parallel to the X axis. Attribute width represents the size of the edges of the box parallel to the Y axis. Attribute height represents the size of the edges of the box parallel to the Z axis. The points of the CuboidFeatureShape are all points on the faces of the box. The box has four faces (the faces with z=0 and z=*length* are missing), so the points are all points represented by the following.
 (-*length*/2, y, z) for –*width*/2 <= y <= *width*/2 and 0 <= z <= *height*
(*length*/2, y, z) for –*width*/2 <= y <= *width*/2 and 0 <= z <= *height*
(x, -*width*/2, z) for –*length*/2 <= x <= *length*/2 and 0 <= z <= *height*
(x, *width*/2, z) for –*length*/2 <= x <= *length*/2 and 0 <= z <= *height*

Class CuboidFeature is a subclass of DirectPlacementFeature used to represent a box shape positioned in a 3D space. CuboidFeature has one additional attribute. Attribute *cuboidShape* represents the shape of the box, and its type is CuboidFeatureShape. The *placement* must be a 3DPlacementZX.

Class RoundCuboidFeatureShape is a subclass of CuboidFeatureShape used to represent a box shape with rounded ends. Specifically, the faces parallel to the YZ plane are replaced by half cylinders that extend outward from the box. The diameters of the half cylinders are equal to the width of the box. The class has all the attributes of CuboidFeatureShape and has no additional attributes. The points of the RoundCuboidFeatureShape are all points represented by the following.

(x, -width/2, z) for -length/2 <= x <= length/2 and 0 <= z <= height
(x, width/2, z) for -length/2 <= x <= length/2 and 0 <= z <= height
((length/2 + R*cos(a)), R*sin(a), z) for $-\pi/2 < a < \pi/2$ and 0 <= z <= height
((-length/2 + R*cos(a)), R*sin(a), z) for $\pi/2 < a < 3\pi/2$ and 0 <= z <= height
where R = width/2.

Class RoundCuboidFeature is a subclass of DirectPlacementFeature used to represent a box with rounded ends shape positioned in a 3D space. RoundCuboidFeature has one additional attribute. Attribute *roundCuboidShape* represents the shape of the round box, and its type is RoundCuboidFeatureShape. The *placement* must be a 3DPlacementZX.

Class EllipticalCylinderFeatureShape is a subclass of 3DSurfaceFeatureShape used to represent a surface that is an elliptical cylinder. EllipticalCylinderFeatureShape has four attributes. Attribute *majorDiameter* represents the major diameter of the ellipse that is the cross section of the feature, and its data type is LengthMeasure. Attribute *minorDiameter* represents the minor diameter of the ellipse that is the cross section of the feature, and its data type is also LengthMeasure. Attribute *insideEllipticalCylinder* represents whether the elliptical cylinder is inside of material or outside of material, and its data type is optional **boolean**. True represents that the elliptical cylinder is inside of material, e.g., the surface of an elliptical cylindrical hole. Attribute *length* represents the length of the elliptical cylinder, and its data type is LengthMeasure. In its native coordinate system, the axis of the elliptical cylinder is the Z axis, the *majorDiameter* is parallel to the X axis, the elliptical cylinder is bounded below by the XY plane, and the elliptical cylinder is bounded above by the plane z = *length*. The points of the elliptical cylinder are all points represented by
 (R*cos(a), r*sin(a), z) for $0 <= a < 2\pi$ and
0 <= z <= length, where R = majorDiameter/2, and r = minorDiameter/2.

Class EllipticalCylinderFeature is a subclass of DirectPlacementFeature used to represent an elliptical cylinder positioned in a 3D space. EllipticalCylinderFeature has one additional attribute. Attribute *ellipticalCylinderShape* represents the shape of the elliptical cylinder, and its type is EllipticalCylinderFeatureShape. The *placement* must be a 3DPlacementZX.

Figure 13 shows the diagram of composite features.

(Figure 13 goes here.)

Class CounteredFeatureShape is an abstract subclass of AxisSymmetricFeature used to represent the surface of a cylindrical hole that is counterbored, countersunk, or counter drilled. CounteredFeatureShape is always inside of material and has two attributes. Attribute *diameter* represents the diameter of the cylinder, and its data type is LengthMeasure. Attribute *length* represents the length of the cylinder, and its data type is LengthMeasure. In its native coordinate system, the cylinder is bounded below by the XY plane and above by the plane z = *length*. The hole may be a blind hole or a through hole. If the hole is a blind hole, it may be extended below the XY plane by a conical surface as would be made by a drill. Any such extension is not represented.

Class CounterboredFeatureShape is a subclass of CounteredFeatureShape used to represent the surface of a cylindrical hole that has been counterbored. The counterbored portion of the hole is a shorter cylindrical surface coaxial with the original hole starting at the top of the original hole and having a larger diameter. The remainder of the original hole is joined to the new cylinder by a planar annulus. CounterboredFeatureShape has two additional attributes. Attribute *counterboreLength* represents the length of the new cylinder, and its data type is LengthMeasure. Attribute *counterboreDiameter* represents the diameter of the new cylinder, and its data type is LengthMeasure. The points of a CounterboredFeatureShape are all points represented by three surfaces.

1. the remaining surface of the original cylinder,
   $(r*cos(a), r*sin(a), z)$, where r = *diameter*/2, $0 <= a < 2\pi$, and
   $0 <= z <= $ *(length – counterboreLength)*
2. the surface of the new cylinder,
   $(R*cos(a), R*sin(a), z)$, where R = *counterboreDiameter*/2, $0 <= a < 2\pi$, and
   *(length – counterboreLength)* $<= z <= $ *length*
3. the surface of the annulus between the two cylinders.

Class CounterboredFeature is a subclass of AxialFeature used to represent a CounterboredFeatureShape positioned in a 3D space. CounterboredFeature has one additional attribute. Attribute *counterboredShape* represents the shape of the CounterboredFeature, and its type is CounterboredFeatureShape. The *placement* must be a 3DPlacementZ or a 3DPlacementZX.

Class CounterDrilledFeatureShape is a subclass of CounteredFeatureShape used to represent the surface of a cylindrical hole that has been counter drilled. The counter drilled portion of the hole is a shorter cylindrical surface coaxial with the original hole starting at the top of the original hole and having a larger diameter. The remainder of the original hole is joined to the new cylinder by a frustum formed by the conical tip of the drill used for counter drilling. The exact shape of the frustum is not represented. CounterDrilledFeatureShape has two additional attributes. Attribute *counterDrillLength* represents the length of the new cylinder, and its data type is LengthMeasure. Attribute *counterDrillDiameter* represents the diameter of the new cylinder, and its data type is LengthMeasure. The points of a CounterDrilledFeatureShape are all points represented by three surfaces.

1. the remaining surface of the original cylinder,
   $(r*cos(a), r*sin(a), z)$, where r = *diameter*/2, $0 <= a < 2\pi$, and
   $0 <= z <= $ *(length – (counterDrillLength* + frustumLength))
2. the surface of the new cylinder,
   $(R*cos(a), R*sin(a), z)$, where R = *counterDrillDiameter*/2, $0 <= a < 2\pi$, and
   *(length – counterDrillLength)* $<= z <= $ *length*
3. the surface of the frustum between the two cylinders.

Class CounterDrilledFeature is a subclass of AxialFeature used to represent a CounterDrilledFeatureShape positioned in a 3D space. CounterDrilledFeature has one additional attribute. Attribute *counterDrilledShape* represents the shape of the CounterDrilledFeature, and its type is CounterDrilledFeatureShape. The *placement* must be a 3DPlacementZ or a 3DPlacementZX.

Class CountersunkFeatureShape is a subclass of CounteredFeatureShape used to represent the surface of a cylindrical hole that has been countersunk. The countersunk portion of the hole is a frustum coaxial with the original hole, having one end joined to the remainder of the original hole by a common circle (whose diameter is that of the original hole), and having a larger diameter at the top of the hole. CountersunkFeatureShape has two additional attributes. Attribute *countersinkAngle* represents the angle of the frustum, and its data type is AngularMeasure. Attribute *countersinkDiameter* represents the larger diameter of the frustum, and its data type is LengthMeasure. The points of a CountersunkFeatureShape are all points represented by two surfaces.

1. the remaining surface of the original cylinder,
   (r*cos(a), r*sin(a), z), where r = *diameter*/2, $0 <= a < 2\pi$, and
   $0 <= z <= $ (*length* – frustumLength)
2. the surface of the frustum,
   (R*cos(a), R*sin(a), z), where R = ((*countersinkDiameter*/2) – ((length – z) * tan(b))),
   $0 <= a < 2\pi$, and (*length* – frustumLength) $<= z <= $ *length*

where b = *countersinkAngle*/2, frustumLength = (*countersinkDiameter* – *diameter*)/(2*tan(b))

Class CountersunkFeature is a subclass of AxialFeature used to represent a CountersunkFeatureShape positioned in a 3D space. CountersunkFeature has one additional attribute. Attribute *countersunkShape* represents the shape of the CountersunkFeature, and its type is CountersunkFeatureShape. The *placement* must be a 3DPlacementZ or a 3DPlacementZX.

Figure 14 shows the diagram of planar features.

(Figure 14 goes here.)

Class PlanarSymmetricFeatureShape is an abstract subclass of 3DSurfaceFeatureShape used to represent a feature that is symmetric about a plane. In the native coordinate system of a PlanarSymmetricFeatureShape, the plane of symmetry is the ZX plane. In mathematical terms, whenever a point (x, y, z) is a point of a PlanarSymmetricFeatureShape, so is (x, -y, z).

Class TwoParallelPlaneFeatureShape is a subclass of PlanarSymmetricFeatureShape used to represent a pair of parallel rectangles. It has four attributes. Attribute *length* represents the size of the rectangles in the X direction, and its data type is LengthMeasure. Attribute *width* represents the size of the rectangles in the Z direction, and its data type is also LengthMeasure. Attribute *gap* represents the distance between the two parallel planes, and its data type is LengthMeasure. Attribute *inside* represents whether the rectangles are inside of material or outside of material, and its data type is optional ***boolean***. True represents that the rectangles are inside of material, e.g., the sides of a slot. In the native coordinate system of the TwoParallelPlaneFeatureShape, the two parallel planes are symmetric about the ZX plane, one pair of edges is parallel to the X axis, the other pair of edges is parallel to the Z axis, and the centers of the rectangles are on the Y axis. The points of the TwoParallelPlaneFeatureShape are all points represented by (x, -*gap*/2, z) or (x, *gap*/2, z) for -*length*/2 $<=$ x $<=$ *length*/2 and -*width*/2 $<=$ z $<=$ *width*/2.

Class TwoParallelPlaneFeature is a subclass of DirectPlacementFeature used to represent a feature that is a pair of parallel rectangles positioned in a 3D space. TwoParallelPlaneFeature has one additional attribute. Attribute *twoParallelPlaneShape* represents the shape of the rectangles, and its type is TwoParallelPlaneFeatureShape. The *placement* must be a 3DPlacementZX.

Class TwoParallelPlaneFlatEndFeatureShape is a subclass of TwoParallelPlaneFeatureShape used to represent a pair of parallel rectangles with a closed flat end at one side of the two parallel rectangles. The closed end is a rectangle whose size in the Z direction is *width* and whose size in the Y direction is *gap*. TwoParallelPlaneFlatEndFeatureShape has all of the attributes of a TwoParallelPlaneFeatureShape plus one additional attribute. Attribute *xPlusEndClosed* represents the closed end of the feature shape, and its data type is **boolean**. True represents that the end where X is positive is closed. False represents that the end where X is negative is closed. A pair of rectangles closed at both ends may be represented using a CuboidFeatureShape. The points of a TwoParallelPlaneFlatEndFeatureShape are all points of a TwoParallelPlaneFeatureShape plus (if *xPlusEndClosed* is true) all points represented by ($length/2$, y, z) for $-gap/2 <= y <= gap/2$ and $-width/2 <= z <= width/2$. If *xPlusEndClosed* is *false*, just put a minus sign in front of *length*/2 in that description.

Class TwoParallelPlaneFlatEndFeature is a subclass of DirectPlacementFeature used to represent a TwoParallelPlaneFlatEndFeatureShape positioned in a 3D space. TwoParallelPlaneFlatEndFeature has one additional attribute. Attribute *twoParallelPlaneFlatEndShape* represents the TwoParallelPlaneFlatEndFeatureShape, and its type is TwoParallelPlaneFlatEndFeatureShape. The *placement* must be a 3DPlacementZX.

Class TwoParallelPlaneRoundEndFeatureShape is a subclass of TwoParallelPlaneFlatEndFeatureShape used to represent a pair of parallel rectangles with a closed end at one side of the two parallel rectangles that is a half cylinder extending away from the rectangles. The length of the half cylinder in the Z direction is *width* and its diameter is *gap*. TwoParallelPlaneRoundEndFeatureShape has all of the attributes of a TwoParallelPlaneFeatureShape plus one additional attribute. Attribute *xPlusEndClosed* represents the closed end of the feature shape, and its data type is **boolean**. True represents that the end where X is positive is closed. False represents that the end where X is negative is closed. A pair of rectangles closed at both ends by half cylinders may be represented using a RoundCuboidFeatureShape. The points of a TwoParallelPlaneRoundEndFeatureShape are all points of a TwoParallelPlaneFeatureShape plus (if *xPlusEndClosed* is true) all points represented by ($length/2 + R*\cos(a)$, $R*\sin(a)$, z) for $-\pi/2 < a < \pi/2$ and $-width/2 <= z <= width/2$, where $R = gap/2$. If *xPlusEndClosed* is false, the additional points are represented by ($-length/2 + R*\cos(a)$, $R*\sin(a)$, z) for $\pi/2 < a < 3\pi/2$ and $-width/2 <= z <= width/2$.

Class TwoParallelPlaneRoundEndFeature is a subclass of DirectPlacementFeature used to represent a TwoParallelPlaneRoundEndFeatureShape positioned in a 3D space. TwoParallelPlaneRoundEndFeature has one additional attribute. Attribute *twoParallelPlaneRoundEndShape* represents the TwoParallelPlaneRoundEndFeatureShape, and its type is TwoParallelPlaneRoundEndFeatureShape. The *placement* must be a 3DPlacementZX.

Figure 15 shows the diagram of 3D-General-Surface features.

(Figure 15 goes here.)

Class GSurfaceFeatureShape is a subclass of 3DSurfaceFeatureShape used to represent a feature that is a general surface. It has one attribute. Attribute *pointsOnSurface* represents the points that are used to define the surface with a specified mathematical algorithm, and its data type is a set of Coordinates3D, with a minimum of 4 elements in the set.

Class GSurfaceFeature is a subclass of DirectPlacementFeature used to represent a feature that is a general 3D surface positioned in a 3D space. GSurfaceFeature has one additional attribute. Attribute *gSurfaceShape* represents the shape of the surface, and its type is GSurfaceFeatureShape. The *placement* must be a 3DPlacementZX.

3.2.3 Indirect Placement Features

Based on simple 3D features, specific form features are defined for applications in design for disassembly and disassembly process planning at the end of a product's useful life. Figure 16 shows the diagram of indirect placement features.

(Figure 16 goes here.)

Class RoundHole is a subclass of IndirectPlacementFeature used to represent a hole in a workpiece. It has one attribute. Attribute *theHole* represents the hole, and its data type is CylinderFeature. The value of the *insideCylinder* attribute of the *cylinderShape* attribute of the CylinderFeature must be *true*.

Class TaperedHole is a subclass of IndirectPlacementFeature used to represent a tapered hole in a workpiece. It has one attribute. Attribute *theHole* represents the tapered hole, and its data type is FrustumFeature. The value of the *insideFrustum* attribute of the *frustumShape* attribute of the FrustumFeature must be *true*.

Class Pin is a subclass of IndirectPlacementFeature used to represent a pin (a protrusion) in a workpiece. It has one attribute. Attribute *thePin* represents the pin, and its data type is CylinderFeature. The value of the *insideCylinder* attribute of the *cylinderShape* attribute of the CylinderFeature must be *false*.

Class TaperedPin is a subclass of IndirectPlacementFeature used to represent a tapered pin in a workpiece. It has one attribute. Attribute *thePin* represents the tapered pin, and its data type is FrustumFeature. The value of the *insideFrustum* attribute of the *frustumShape* attribute of the FrustumFeature must be *false*.

Class Slot is a subclass of IndirectPlacementFeature used to represent the parent type for slots, which are depressions with parallel sides and a flat bottom. It has two attributes. Attribute *basePlane* represents the base plane from which the slot is depressed, and its data type is DatumPlane. (DatumPlane is defined in Section 3.3.1.) Attribute *filletRadius* is optional and represents the radius of fillets applied to the concave edges between the bottom of the slot and the sides of the slot. Its data type is LengthMeasure. The *filletRadius* must not exceed either the depth of the slot or half of the width of the slot. The two parallel planes that form the sides of the slot are represented in the subclasses of Slot.

Class FlatEndsClosedSlot is a subclass of Slot used to represent a slot closed at both ends by planes. It has one additional attribute. Attribute *sidesAndEnds* represents the sides and ends of the closed end slot, and its data type is CuboidFeature. The CuboidFeature must be placed so that its top edges are on the *basePlane*. The length, width, and depth of the slot are determined by the length, width, and height of the *cuboidShape* of the CuboidFeature. The *insideCuboid* attribute of the *cuboidShape* of the CuboidFeature must be *true*.

Class RoundEndsClosedSlot is a subclass of Slot used to represent a slot closed at both ends by half cylinders. It has one additional attribute. Attribute *sidesAndEnds* represents the sides and ends of the closed end slot, and its data type is RoundCuboidFeature. The RoundCuboidFeature must be placed so that its top edges are on the *basePlane*. The length, width, and depth of the slot are determined by the length, width, and height of the *roundCuboidShape* of the RoundCuboidFeature. The *insideCuboid* attribute of the *roundCuboidShape* of the RoundCuboidFeature must be *true*.

Class OneRoundEndSlot is a subclass of Slot used to represent a slot closed at one end by a half cylinder. It has one additional attribute. Attribute *sidesAndEnd* represents the sides and end of the slot, and its data type is TwoParallelPlaneRoundEndFeature. The TwoParallelPlaneRoundEndFeature must be placed so that its top edges are on the *basePlane*. The length, width, and depth of the slot are determined by the length, gap, and width of the *twoParallelPlaneRoundEndShape* of the TwoParallelPlaneRoundEndFeature. The *inside* attribute of the *twoParallelPlaneRoundEndShape* of the TwoParallelPlaneRoundEndFeature must be *true*.

Class OneFlatEndSlot is a subclass of Slot used to represent a slot closed at one end by a plane. It has one additional attribute. Attribute *sidesAndEnd* represents the sides and end of the slot, and its data type is TwoParallelPlaneFlatEndFeature. The TwoParallelPlaneFlatEndFeature must be placed so that its top edges are on the *basePlane*. The length, width, and depth of the slot are determined by the length, gap, and width of the *twoParallelPlaneFlatEndShape* of the TwoParallelPlaneFlatEndFeature. The *inside* attribute of the *twoParallelPlaneFlatEndShape* of the TwoParallelPlaneFlatEndFeature must be *true*.

Class OpenEndsSlot is a subclass of Slot used to represent a slot open at both ends. It has one additional attribute. Attribute *sides* represents the sides of the slot, and its data type is TwoParallelPlaneFeature. The TwoParallelPlaneFeature must be placed so that its top edges are on the *basePlane*. The length, width, and depth of the slot are determined by the length, gap, and width of the *twoParallelPlaneShape* of the TwoParallelPlaneFeature. The *inside* attribute of the *twoParallelPlaneShape* of the TwoParallelPlaneFeature must be *true*.

Class Rib is a subclass of IndirectPlacementFeature used to represent a rib, a protrusion. It has two attributes. Attribute *basePlane* represents the base plane from which the rib protrudes, and its data type is DatumPlane. Attribute *filletRadius* is optional and represents the radius of fillets applied to the concave edges between the *basePlane* and the sides of the rib. Its data type is LengthMeasure. The *filletRadius* must not exceed the height of the rib. The ends of the rib and the two parallel planes that form the sides of the rib are represented in the subclasses of Rib.

Class RoundEndsRib is a subclass of Rib used to represent a rib with rounded ends. It has one additional attribute. Attribute *sidesAndEnds* represents the sides and ends of the rib, and its data type is RoundCuboidFeature. The RoundCuboidFeature must be placed so that its bottom edges are on the *basePlane*. The length, width, and height of the rib are determined by the length, width, and height of the *roundCuboidShape* of the RoundCuboidFeature. The *insideCuboid* attribute of the *roundCuboidShape* of the RoundCuboidFeature must be *false*.

Class FlatEndsRib is a subclass of Rib used to represent a rib with flat ends. It has one additional attribute. Attribute *sidesAndEnds* represents the sides and ends of the rib, and its data type is CuboidFeature. The CuboidFeature must be placed so that its bottom edges are on the *basePlane*. The length, width, and height of the rib are determined by the length, width, and height of the *cuboidShape* of the CuboidFeature. The *insideCuboid* attribute of the *cuboidShape* of the CuboidFeature must be *false*.

Class RectangularBoss is a subclass of IndirectPlacementFeature used to represent a boss (a protrusion) on a workpiece. A RectangularBoss has two sets of parallel sides and a flat top. Adjacent sides are orthogonal to each other. It has four attributes. Attribute *basePlane* represents the base plane from which the boss protrudes, and its data type is Plane. Attribute *filletRadius* is optional and represents the radius of fillets applied to the concave edges between the bottom edges of the boss and the *basePlane*. Its data type is LengthMeasure. The *filletRadius* must not exceed the height of the boss. Attribute *sidesAndEnds* represents the sides and ends of the pocket, and its data type is CuboidFeature. The CuboidFeature must be placed so that its bottom edges are on the *basePlane*. The length, width, and height of the boss are determined by the length, width, and height of the *cuboidShape* of the CuboidFeature. The *insideCuboid* attribute of the *cuboidShape* of the CuboidFeature must be false. Attribute *cornerRadius* represents rounding the four edges of the boss that do not lie on the top by removing material. Its data type is LengthMeasure. The *cornerRadius* must not exceed half of the width of the boss.

Class RectangularPocket is a subclass of RectangularBossFeature used to represent a pocket (a depression) in a workpiece. A RectangularPocket has two sets of parallel sides and a flat bottom. Adjacent sides are orthogonal to each other. It has four attributes. Attribute *basePlane* represents the base plane from which the pocket is depressed, and its data type is Plane. Attribute *filletRadius* is optional and represents the radius of fillets applied to the concave edges between the bottom of the pocket and the sides of the pocket. Its data type is LengthMeasure. The *filletRadius* must not exceed either the depth of the pocket or half of the width of the pocket. Attribute *sidesAndEnds* represents the sides and ends of the pocket, and its data type is CuboidFeature. The CuboidFeature must be placed so that its top edges are on the *basePlane*. The length, width, and depth of the pocket are determined by the length, width, and height of the *cuboidShape* of the CuboidFeature. The *insideCuboid* attribute of the *cuboidShape* of the CuboidFeature must be true. Attribute *cornerRadius* represents rounding the corners of the pocket (by adding material). Its data type is LengthMeasure. The *cornerRadius* must not exceed half of the width of the pocket.

3.2.4 Implicit Placement Features

Implicit placement features are defined for generating inspection features. Figure 17 shows the diagram of all the implicit placement features classes. As mentioned earlier, an ImplicitPlacementFeature modifies one or more other features. The placement of an ImplicitPlacementFeature is determined by the placement(s) of the modified feature(s) and the nature of the modification.

(Figure 17 goes here.)

Class Chamfer is a subclass of ImplicitPlacementFeature used to represent the removal of material from a convex edge on a workpiece in such a way that a new surface and two new edges are created. The new surface created is flat in at least one direction in the sense that is possible to draw line segments on it that meet the new edges (or the tangents to the new edges) in a right angle. For example, if the original edge is formed by two planes, the new surface is a plane. As another example, if the original edge is formed by a plane and a cylinder, and the plane is orthogonal to the axis of the cylinder, the new surface is a frustum of a cone. In general, imagine that the material is cheese and the chamfer is cut by a wire cheese cutter moving along the original edge so that a narrow long slice is removed that includes the original edge. Chamfer has four attributes. Attribute *firstReferenceSurface* represents the first reference surface in creating a chamfer, and its data type is DirectPlacementFeature, with the restriction that the FeatureShape of the DirectPlacementFeature must be a surface. Attribute *firstOffset* represents the distance from the original edge to the new edge on the first reference surface, and its data type is LengthMeasure. Attribute *secondReferenceSurface* represents the second reference surface in creating a chamfer, and its data type is DirectPlacementFeature, with the restriction that the FeatureShape of the DirectPlacementFeature must be a surface. Attribute *secondOffset* represents the distance from the original edge to the new edge on the second reference surface, and its data type is LengthMeasure.

Class Fillet is a subclass of ImplicitPlacementFeature used to represent a fillet of a concave edge between two surfaces of a workpiece. A fillet adds material at the edge to which it is applied. The shape of the fillet is what would be created by filling the edge with putty and then smoothing it with a rolling ball that stays in contact with the surfaces that form the edge. Fillet has three attributes. Attribute *firstReferenceSurface* represents the first reference surface in creating a fillet, and its data type is DirectPlacementFeature, with the restriction that the FeatureShape of the DirectPlacementFeature must be a surface. Attribute *secondReferenceSurface* represents the second reference surface in creating a fillet, and its data type is DirectPlacementFeature, with the restriction that the FeatureShape of the DirectPlacementFeature must be a surface. Attribute *radius* represents the radius of the fillet (i.e., the radius of the imaginary rolling ball), and its data type is LengthMeasure.

Class RoundEdge is a subclass of Fillet used to represent the removal of material from a convex edge of a workpiece by rounding the edge. Specifically, at each point on the edge, the lines of intersection of the two surfaces that form the edge with a plane perpendicular to the edge (or perpendicular to the tangent to the edge) are joined by a circular arc that is tangent to the lines.

RoundEdge has three attributes. Attribute *firstReferenceSurface* represents the first reference surface in creating a round edge, and its data type is DirectPlacementFeature, with the restriction that the FeatureShape of the DirectPlacementFeature must be a surface. Attribute *secondReferenceSurface* represents the second reference surface in creating a round edge, and its data type is DirectPlacementFeature, with the restriction that the FeatureShape of the DirectPlacementFeature must be a surface. Attribute *radius* represents the radius of the round edge, and its data type is LengthMeasure.

Class RoundedCorner is a subclass of ImplicitPlacementFeature used to represent the removal of material from a convex corner of a workpiece by rounding the corner. The surface of the rounded corner is part of the surface of a sphere. RoundedCorner has two attributes. Attribute *corner* represents the corner that is rounded, and its data type is PointFeature. Attribute *radius* represents the radius of the round corner, and its data type is LengthMeasure.

Special features are defined for generating features with special purposes, such as thread and screw. More special features, such as gear tooth and helical spring, can be added when they are needed.

Class GeneralThread is used to represent a general thread. It has eight attributes. Attribute *fitClass* represents how tight the assembled threads are, and its data type is String. Fit class is defined in ISO 10303-Part 224. Attribute *form* represents the geometric shape of the thread, and its data type is String. Attribute *innerThread* represents whether the thread is an inner thread. If it is not, it is an outer thread. The data type is **boolean**. A value of true means the thread is an inner thread. Attribute *majorDiameter* represents the major diameter of the thread, and its data type is LengthMeasure. Attribute *minorDiameter* represents the minor diameter of the thread, and its data type is LengthMeasure. Attribute *pitchDiameter* represents the pitch diameter of the thread, and its data type is LengthMeasure. Attribute *numberOfThread* represents the number of threads in a unit of length, and its data type is **int**. Attribute *rightHanded* represents whether the thread is right handed in orientation. If it is not, it is left handed. The data type is **boolean**. A value of true means the thread is right handed.

Class ThreadedHole is a subclass of ImplicitPlacementFeature used to represent a threaded hole in a workpiece. ThreadedHole has two attributes. Attribute *theThread* represents the thread of a hole, and its data type is GeneralThread. Attribute *theHole* represents the hole, and its data type is RoundHole. The *innerThread* of *theThread* must be *true*.

Class Screw is a subclass of ImplicitPlacementFeature used to represent a threaded pin on a workpiece. It has two attributes. Attribute *theThread* represents the thread of a pin, and its data type is GeneralThread. Attribute *thePin* represents the pin and its data type is Pin. The *innerThread* of *theThread* must be *false*.

Class CrestedThread is a subclass of GeneralThread used to represent a thread with a crest. It has one attribute. Attribute *crest* represents the dimension of the crest, and its data type is LengthMeasure.

3.2.5 Pattern Shapes and Features

Figure 18 shows the diagram of matrix pattern shapes and features.

(Figure 18 goes here.)

Class PatternFeatureShape is a subclass of FeatureShape used to represent a pattern of feature shapes. Figure 11 shows the diagram of all the pattern feature classes in the PatternFeaturePack package. Pattern feature is defined in ANSI/ASME Y14.5 standard [26]. PatternFeatureShape has three attributes. Attribute *baseFeatureShape* represents a reference feature for all the features in the pattern, and its data type is FeatureShape (excluding PatternFeatureShape). Attribute *zDirection* represents the direction in the native coordinate system of the pattern of the Z axis of the native coordinate system of the *baseFeatureShape*, and is data type is is UnitVector3D. Attribute *xDirection* represents the direction in the native coordinate system of the pattern of the X axis of the native coordinate system of the *baseFeatureShape*, and is data type is UnitVector3D.

Each subclass of PatternFeatureShape says how to locate copies of the *baseFeatureShape* in the native coordinate system of the PatternFeatureShape. Locating a FeatureShape is done using a point of the FeatureShape as a location point. The location point of a FeatureShape that is a depression (e.g., a hole) is the highest point of the FeatureShape on the Z axis of the native coordinate system of the FeatureShape. The location point of a FeatureShape that is a protrusion (e.g., a boss) is the origin of the native coordinate system of the FeatureShape (the origin is the lowest point of the FeatureShape).

Class MatrixPatternFeatureShape is a subclass of PatternFeatureShape used to represent a planar array of feature shapes. The array has rows and columns in which feature shapes are evenly spaced. Elements of rows and columns are indexed. The first element of a row or column has index 1. The class has four additional attributes. Attribute *numberOfFeaturesInRow* represents the number of features in a row of the pattern, and its data type is PositiveInteger. Attribute *numberOfFeaturesInColumn* represents the number of features in a column of the pattern, and its data type is PositiveInteger. The first row of the pattern lies on positive X axis of the native coordinate system of the pattern. Other rows are parallel to the first row. Attribute *columnDirection* represents the column direction of the pattern, and its data type is AngleMeasure. The angle is 0 on the positive X axis and it increases counterclockwise as viewed from the positive Z axis. Attribute *rowInterval* represents the distance between any two features in the row direction, and its data type is LengthMeasure. Attribute *columnInterval* represents the distance between any two features in the column direction of the pattern, and its data type is LengthMeasure. The pattern element whose row and column indices are both 1 has its location point at the origin of the native coordinate system of the pattern. The location points of the rest of the pattern elements are at the points of the array. All pattern elements are oriented as specified by the *zDirection* and the *xDirection*. If *numberOfFeaturesInColumn* is 1, there is a single row of features, and the value of *columnInterval* is not used in determining the placement of the features. If *numberOfFeaturesInRow* is 1, there is a single column of features, and the value of *rowInterval* is not used in determining the placement of the features.

Class MatrixPatternFeature is a subclass of DirectPlacementFeature used to represent an array pattern positioned in a 3D space. MatrixPatternFeature has one additional attribute. Attribute *matrixPatternShape* represents the shape of the array pattern, and its type is MatrixPatternFeatureShape. The *placement* must be a 3DPlacementZX and must place the array pattern on a plane.

Class MatrixOmittedPatternFeatureShape is a subclass of MatrixPatternFeatureShape used to represent an array of feature shapes with some feature shapes that are omitted. The class has two additional attributes. Attribute *rowIndices* represents indices of the omitted features in the row direction of the pattern, and its data type is a list of PositiveInteger. Attribute *columnIndices* represents indices of the omitted features in the column direction of the pattern, and its data type is a list of PositiveInteger. The two lists must be the same length. Corresponding indices in the two lists (i.e., the nth element of each list) are the row and column indices of an element to omit from the pattern. No two pairs of corresponding indices may be the same (i.e., the same pattern element cannot be omitted more than once).

Class MatrixOmittedPatternFeature is a subclass of DirectPlacementFeature used to represent an array pattern with some elements omitted positioned in a 3D space. MatrixOmittedPatternFeature has one additional attribute. Attribute *matrixOmittedPatternShape* represents the shape of the array pattern, and its type is MatrixOmittedPatternFeatureShape. The *placement* must be a 3DPlacementZX and must place the array pattern on a plane.

Class MatrixModifiedPatternFeatureShape is a subclass of MatrixOmittedPatternFeatureShape used to represent a matrix pattern of features with zero to many features that are offset from their normal positions and zero to many features that are omitted. The class has two additional attributes. Attribute *offsetDirections* represents offset directions of the offset features, and its data type is a list of UnitVector3D. The Z coordinates of the *offsetDirections* must be 0. Attribute *offsetDistances* represents offset distances of the offset features, and its data type is a list of LengthMeasure. The lengths of the two lists must be the same as the lengths of the *rowIndices* and *columnIndices* lists. If an element of the *offsetDistances* list is 0, the indicated array element should be omitted (and the corresponding value of *offsetDirections* is not used). If an element of the *offsetDistances* list is positive, the indicated array element should be moved from its array location by that distance in the direction specified by the corresponding element of the *offsetDirections*.

Class MatrixModifedPatternFeature is a subclass of DirectPlacementFeature used to represent an array pattern with some elements omitted or offset positioned in a 3D space. MatrixModifedPatternFeature has one additional attribute. Attribute *matrixModifiedPatternShape* represents the shape of the array pattern, and its type is MatrixModifiedPatternFeatureShape. The *placement* must be a 3DPlacementZX and must place the array pattern on a plane.

Figure 19 shows the diagram of circular pattern shapes and features.

(Figure 19 goes here.)

31

Class CircularPatternFeatureShape is a subclass of PatternFeatureShape used to represent a circular arc pattern of feature shapes. Elements of the pattern are indexed. The first element of the pattern has index 1. The class has four additional attributes. Attribute *numberOfFeatures* represents the number of feature shapes in the pattern, and its data type is PositiveInteger. Attribute *patternRadius* represents the radius of the pattern, and its data type is LengthMeasure. Attribute *peripheralInterval* represents the interval between any two features in the tangential direction of the pattern of features, and its data type is AngularMeasure. Angles are positive in the counterclockwise direction as viewed from the positive Z axis. Attribute *rotateOrientation* represents whether the orientation of pattern elements should be rotated, and its data type is **boolean**. If *rotateOrientation* is false, the native coordinate system of each pattern element should be oriented in the native coordinate system of the pattern as specified by *zDirection* and *xDirection*. If *rotateOrientation* is true, the first element of the pattern should be oriented as specified by *zDirection* and *xDirection*, but other elements should have their orientation rotated (about a vertical axis through the location point) through the angle [(n-1)*peripheralInterval], where n is the index of the element. This is the same as the angle through which the location point has been rotated. The center of the pattern is at the origin of the native coordinate system of the CircularPatternFeatureShape. The first element of the pattern is located on the X axis at (*patternRadius*, 0, 0) in the native coordinate system of the CircularPatternFeatureShape.

Class CircularPatternFeature is a subclass of DirectPlacementFeature used to represent a circular arc pattern positioned in a 3D space. CircularPatternFeature has one additional attribute. Attribute *circularPatternShape* represents the shape of the circular arc pattern, and its type is CircularPatternFeatureShape. The *placement* must be a 3DPlacementZX and must place the circular arc pattern on a plane.

Class CircularOmittedPatternFeatureShape is a subclass of CircularPatternFeatureShape used to represent a circular arc pattern of features with some features that are omitted. The class has one additional attribute. Attribute *indices* represents indices of the omitted features in the pattern, and its data type is a set of PositiveInteger.

Class CircularOmittedPatternFeature is a subclass of DirectPlacementFeature used to represent a circular arc pattern with some elements omitted positioned in a 3D space. CircularOmittedPatternFeature has one additional attribute. Attribute *circularOmittedPatternShape* represents the shape of the circular arc pattern, and its type is CircularOmittedPatternFeatureShape. The *placement* must be a 3DPlacementZX and must place the circular arc pattern on a plane.

Class CircularModifiedPatternFeatureShape is a subclass of Class CircularOmittedPatternFeatureShape used to represent a circular pattern of features with zero to many elements that are offset from their normal positions and zero to many elements that are omitted. The class has two attributes. Attribute *offsetDirections* represents offset directions of the offset features, and its data type is a list of UnitVector3D. The Z coordinates of the *offsetDirections* must be 0. Attribute *offsetDistances* represents offset distances of the offset features, and its data type is a list of LengthMeasure. The lengths of the two lists must be the same as the length of the *indices* list. If an element of the *offsetDistances* list is 0, the corresponding pattern element should be omitted (and the corresponding value of *offsetDirections* is not used). If an element of the *offsetDistances* list is positive, the corresponding pattern element should be moved from its circular arc location by that distance in the direction specified by the corresponding element of the *offsetDirections*. If *rotateOrientation* is true and an element is offset, the element should be rotated from the orientation specified by *zDirection* and *xDirection* by an angle that is the same as the angle between the X axis of the native coordinate system of the pattern and the line from the origin to the location point of the element.

Class CircularModifiedPatternFeature is a subclass of DirectPlacementFeature used to represent a circular arc pattern with some elements omitted or offset positioned in a 3D space. CircularModifiedPatternFeature has one additional attribute. Attribute *circularModifiedPatternShape* represents the shape of the circular arc pattern, and its type is CircularModifiedPatternFeatureShape. The *placement* must be a 3DPlacementZX and must place the circular arc pattern on a plane.

Class ConcentricCircularPatternFeatureShape is a subclass of CircularPatternFeatureShape used to represent a pattern that is set of concentric circular arcs of feature shapes. The class has two additional attributes. Attribute *numberOfFeaturesInRadialDirection* represents the number of elements in the radial direction, and its data type is PositiveInteger. Attribute *radialInterval* represents the interval between adjacent arcs of elements in the radial direction, and its data type is LengthMeasure. The first element of each arc is located on the X axis of the native coordinate system of the pattern. In each arc, the number of elements and the angular separation of adjacent elements is given by the *numberOfFeatures* and *peripheralInterval* inherited from CircularPatternFeatureShape. Elements of the pattern are indexed by an angular index and a radial index. The radial index of each element of the smallest arc is 1. The radial index of each element of the next arc is 2, and so on. The angular index of the first element in each arc is 1 and increases around the arc.

Class ConcentricCircularPatternFeature is a subclass of DirectPlacementFeature used to represent a concentric circular arc pattern positioned in a 3D space. ConcentricCircularPatternFeature has one additional attribute. Attribute *concentricCircularPatternShape* represents the shape of the concentric circular arc pattern, and its type is ConcentricCircularPatternFeatureShape. The *placement* must be a 3DPlacementZX and must place the concentric circular arc pattern on a plane.

Class ConcentricCircularOmittedPatternFeatureShape is a subclass of ConcentricCircularPatternFeatureShape used to represent a concentric circular arc pattern of feature shapes with some feature shapes that are omitted. The class has two additional attributes. Attribute *angularIndices* represents indices of the omitted features in the angular direction, and its data type is a list of PositiveInteger. Attribute *radialIndices* represents indices of the omitted features in the radial direction, and its data type is a list of PositiveInteger. The two lists must be the same length. Corresponding indices in the two lists (i.e., the nth element of each list) are the angular and radial indices of an element to omit from the pattern. No two pairs of corresponding indices may be the same (i.e., the same pattern element cannot be omitted more than once).

33

Class ConcentricCircularOmittedPatternFeature is a subclass of DirectPlacementFeature used to represent a concentric circular arc pattern with some elements omitted positioned in a 3D space. ConcentricCircularOmittedPatternFeature has one additional attribute. Attribute *concentricCircularOmittedPatternShape* represents the shape of the concentric circular arc pattern, and its type is ConcentricCircularOmittedPatternFeatureShape. The *placement* must be a 3DPlacementZX and must place the concentric circular arc pattern on a plane.

Class ConcentricCircularModifiedPatternFeatureShape is a subclass of ConcentricCircularOmittedPatternFeatureShape used to represent a concentric circular arc pattern of features with zero to many features that are offset from their normal positions and zero to many features that are omitted. The class has two additional attributes. Attribute *offsetDirections* represents offset directions of the offset features, and its data type is a list of UnitVector3D. The Z coordinate of each of the *offsetDirections* must be 0. Attribute *offsetDistances* represents offset distances of the offset features, and its data type is a list of LengthMeasure. The lengths of the two lists must be the same as the lengths of the *angularIndices* and *radialIndices* lists. If an element of the *offsetDistances* list is 0, the element should be omitted (and the corresponding value of *offsetDirections* is not used). If an element of the *offsetDistances* list is positive, the indicated pattern element should be moved from its concentric circular arc location by that distance in the direction specified by the corresponding element of the *offsetDirections*. If *rotateOrientation* is true and an element is offset, the element should be rotated from the orientation specified by *zDirection* and *xDirection* by an angle that is the same as the angle between the X axis of the native coordinate system of the pattern and the line from the origin to the location point of the element.

Class ConcentricCircularModifedPatternFeature is a subclass of DirectPlacementFeature used to represent a concentric circular arc pattern with some elements omitted or offset positioned in a 3D space. ConcentricCircularModifiedPatternFeature has one additional attribute. Attribute *concentricCircularModifiedPatternShape* represents the shape of the concentric circular arc pattern, and its type is ConcentricCircularModifedPatternFeatureShape. The *placement* must be a 3DPlacementZX and must place the concentric circular arc pattern on a plane.

3.2.6 Separator

Enumeration type SeparatorType is used to define the type of a separator so that proper disassembly tools can be selected. This enumeration type includes *Bolt-Nut-Washer*, *Screw*, *TaperFit*, *PinFit*, *Rivet*, *SplineFit*, *RubberRing*, *SpringFit*, *Bearing*, *GearMesh*, *BeltMesh*, *Glue*, and *Welding* [24].

Class Separator is a subclass of Class Connector, defined in the Open Assembly Model, and is used to represent a separator in an assembly that is to be separated at the end of its useful life. It has two attributes. Attribute *feature* represents the feature from which the assembly will be separated into subassemblies, and its data type is SimpleFeature. Attribute *type* represents the type of the separator, and its data type is SeparatorType.

**4. Conclusion and Future Work**

Manufacturing industries are facing the challenge of reuse and recycle of products at the end of the products' service lives. Our literature review shows that the number of environmental protection regulations is increasing rapidly. Reuse and recycle are critical activities to alleviate natural resource depletion and save energy to achieve the goal of sustainable development. Disassembly of out-of-service products is a key operation to separate the product into reusable and recyclable parts. Information on design for disassembly and disassembly process planning is critical for decision making in design and manufacturing. An information model for disassembly processes is, hence, developed using the Unified Modeling Language. The model includes shape, placement, feature composition classes and their relationships that define disassembly features. The developed information model provides a basis for software development of design for disassembly and disassembly process planning systems.

Possible future work includes comprehensive tests of the information model with more complicated designs. These are needed to test the model. Prototype disassembly databases, cost of disassembly analysis software, and disassembly process planning systems can be developed using the information model. Finally, a standard data exchange format of the XML (eXtensible Markup Language) model for design for disassembly and disassembly process plans can also be developed.

## References

1　Rumbaugh, J., Jacobson, I., and Booch, G., The Unified Modeling Language Reference Manual, 2nd edition, Addison Wesley, 2004.

2　Vinodh, S., Praveen Kumar, R., and Nachiappan, N., "Disassembly modeling, planning, and leveling for a cam-operated rotary switch assembly: A case study," International Journal of Advanced Manufacturing Technology, Vol. 62, pp. 789-800, 2012.

3　Tang, Y., Zhou, M., and Caudill, R., "An integrated approach to disassembly planning and demanufacturing operation," IEEE Transactions on Robotics and Automation, 17, pp. 773-784, 2001.

4　Mascle, C. and Zhao, H., "Integrating environmental consciousness in product/process development based on life-cycle thinking," International Journal of Production Economics, 112, pp. 5-17, 2008.

5　Behdad, S. and Thurston, D., "Disassembly and reassembly sequence planning tradeoffs under uncertainty for product maintenance," Transactions of the ASME Journal of Mechanical Design, Vol. 134, pp. 041011-1 to 041011-9, 2012.

6　Fan, S.-K.. Fan, C.. Yang, J.-H., and Liu, K., "Disassembly and recycling cost analysis of waste notebook and the efficiency improvement by re-design process," Journal of Cleaner Production, Vol. 39, pp. 209-219, 2013.

7　Liu, X., Peng, G., Liu, X., and Hou, Y., "Disassembly sequence planning approach for product virtual maintenance based on improved max-min ant system," International Journal of Advanced Manufacturing Technology, Vol. 59, pp. 829-839, 2012.

8　Smith, S., Smith, G., and Chen, W.-H., "Disassembly sequence structure graphs: An optimal approach for multiple-target selective disassembly sequence planning," Journal of Advanced Engineering Informatics, Vol. 26, pp. 306-316, 2012.

9    Tseng, H.-E., Chang, C.-C., and Cheng, C.-J., "Disassembly-oriented assessment methodology for product modularity," International Journal of Production Research, Vol. 48, pp. 4297-4320, 2010.

10   Li, J., Wang, Q., and Huang, P., "An integrated disassembly constraint generation approach for product design evaluation," International Journal of Computer Integrated Manufacturing, Vol. 25, pp. 565-577, 2012.

11   ISO 10303-44:2000, Industrial Automation Systems and Integration - Product Data Representation and Exchange - Part 44: Integrated Generic Resources: Product Structure Configuration, 2nd edition.

12   Sudarsan, R., Han, Y., Foufou, S., Feng, S., Roy, U., Wang, F., Sriram, R., and Lyons, K., "A Model for Capturing Product Assembly Information," Transactions of ASME, Journal of Computing and Information Science in Engineering, Vol. 6, March 2006, pp. 11 – 21

13   Vinodh, S., Nachiappan, N., and Praveen Kumar, R., "Sustainability through disassembly modeling, planning, and leveling: A case study," Journal of Clean Technologies and Environmental Policy, Vol. 14, pp. 55-67, 2012.

14   Lambert, A. and Gupta, S., "Demand-driven disassembly optimization for electronic products package reliability," Journal of Electronics Manufacturing, Vol. 11, pp. 121-135, 2002.

15   Murshed, S., Dixon, A., and Shah, J., "Neutral Definition and Recognition of Assembly Features for Legacy Systems Reverse Engineering", ASME 2009 International Design Engineering Technical Conferences and Computers and Information in Engineering Conference (IDETC/CIE2009)  August 30 – September 2, 2009 , San Diego, California, USA, Paper No DETC2009-86739.

16   Murshed, M., Shah J., and Jagasivamani, V., "OAM+: An Assembly data model for Legacy Systems Engineering", ASME 2007 International Design Engineering Technical Conferences and Computers and Information in Engineering Conference (IDETC/CIE2007) September 4 – 7, 2007 , Las Vegas, Nevada, USA, Paper No: DETC2007-35723.

17   Chan, C. and Tan, S., "Generating assembly features onto split solid models," Computer-Aided Design, Vol.35, pp.1315 - 1336, 2003.

18   Hamidullah, B. and Irfan, M., "Assembly features: definition, classification, and instantiation," IEEE-ICET 2006 2nd International Conference on Emerging Technologies, Preshawar, Pakistan, pp.617-623, November 2006.

19   ISO 10303-224:2006, Industrial automation systems and integration - Product data representation and exchange - Part 224: Application protocol: Mechanical product definition for process planning using machining features.

20   ISO 10303-111:2007, Industrial automation systems and integration - Product data representation and exchange - Part 111: Integrated application resource: Elements for the procedural modelling of solid shapes.

21   Dipper, T., Xu, X., and Klemm, P., "Defining, recognizing and representing feature interactions in a feature-based data model," Journal of Robotics and Computer-Integrated Manufacturing, Vol. 27, pp. 101-114, 2011.

22   ISO 10303-238:2007, Industrial automation systems and integration - Product data representation and exchange - Part 238: Application protocol: Application interpreted model for computerized numerical controllers.

23  Hu, Y., Wang, Y., Zhao, G., Wang, Y., and Yuan, X., "Feature-based modeling of automobile gears and manufacturing resources for virtual manufacturing," International. Journal of Advanced Manufacturing Technology, Vol. 55, pp. 405-419, 2011.

24  Anjum, N., Harding, J., and Young, R., and Case, K., "Manufacturability verification through feature-based ontological product models, Proceedings of the Institution of Mechanical Engineers, Part B: Journal of Engineering Manufacture, Vol. 226, pp. 1086-1098, 2012.

25  ISO 10303-203:2011, Industrial automation systems and integration – Product data representation and exchange – Part 203: Application protocol: Configuration Controlled 3D Designs of Mechanical Parts and Assemblies.

26  ANSI/ASME Y14.5, Dimensioning and Tolerancing, The American Society of Mechanical Engineers, New York City, New York, 2009.

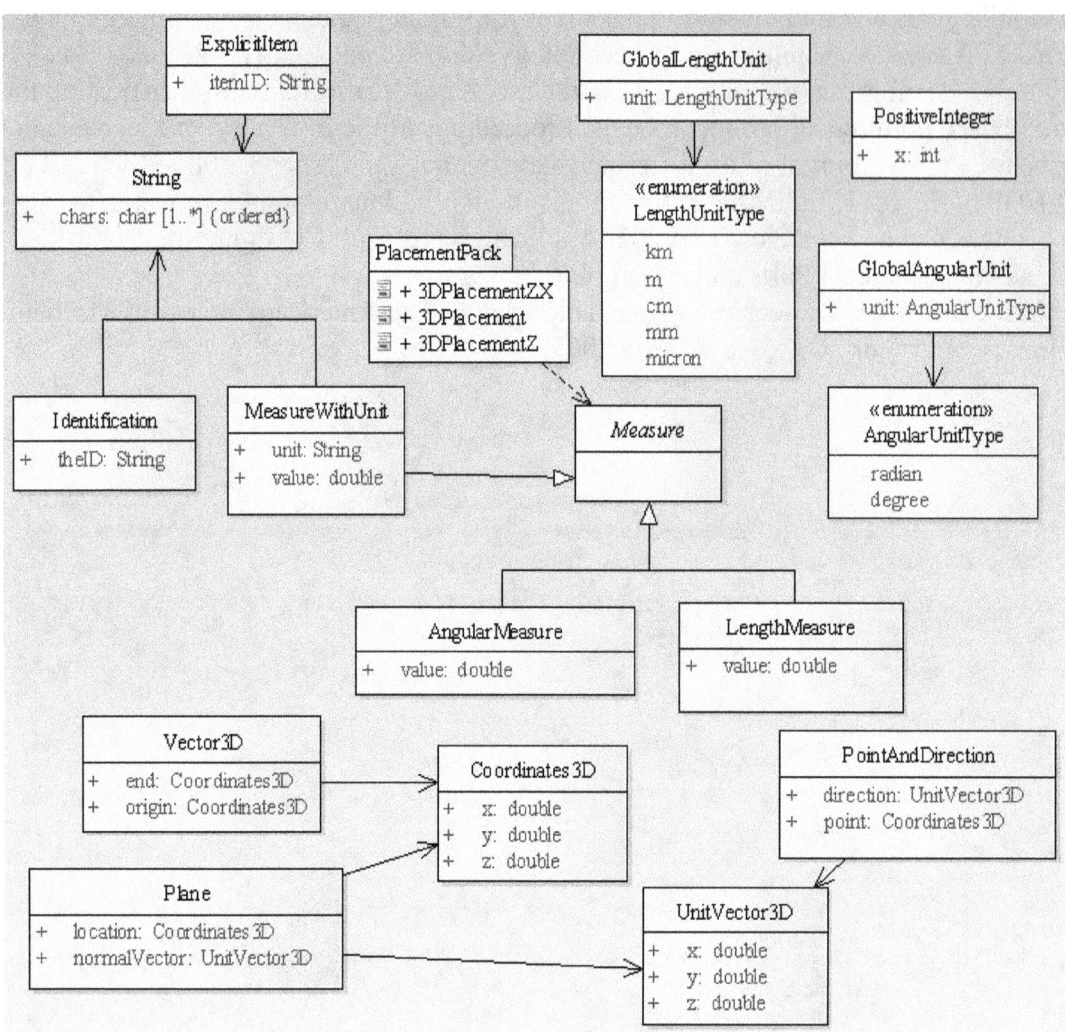

Figure 1. Class diagram of Support Data

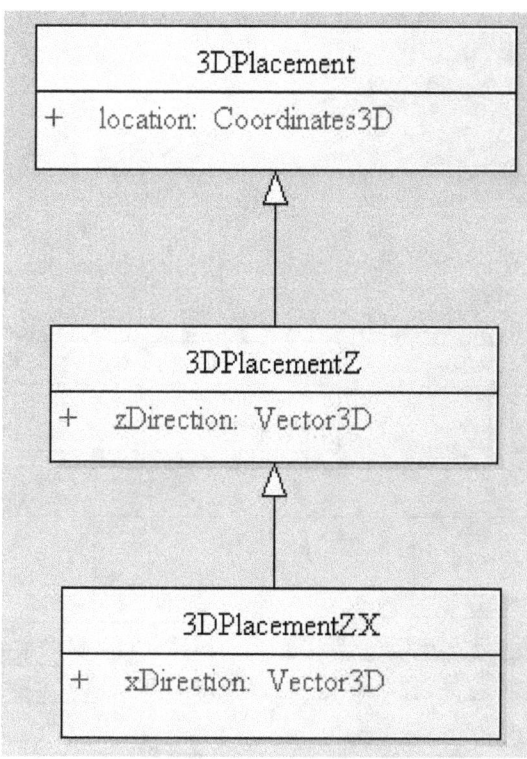

Figure 2. Class diagram of Placement

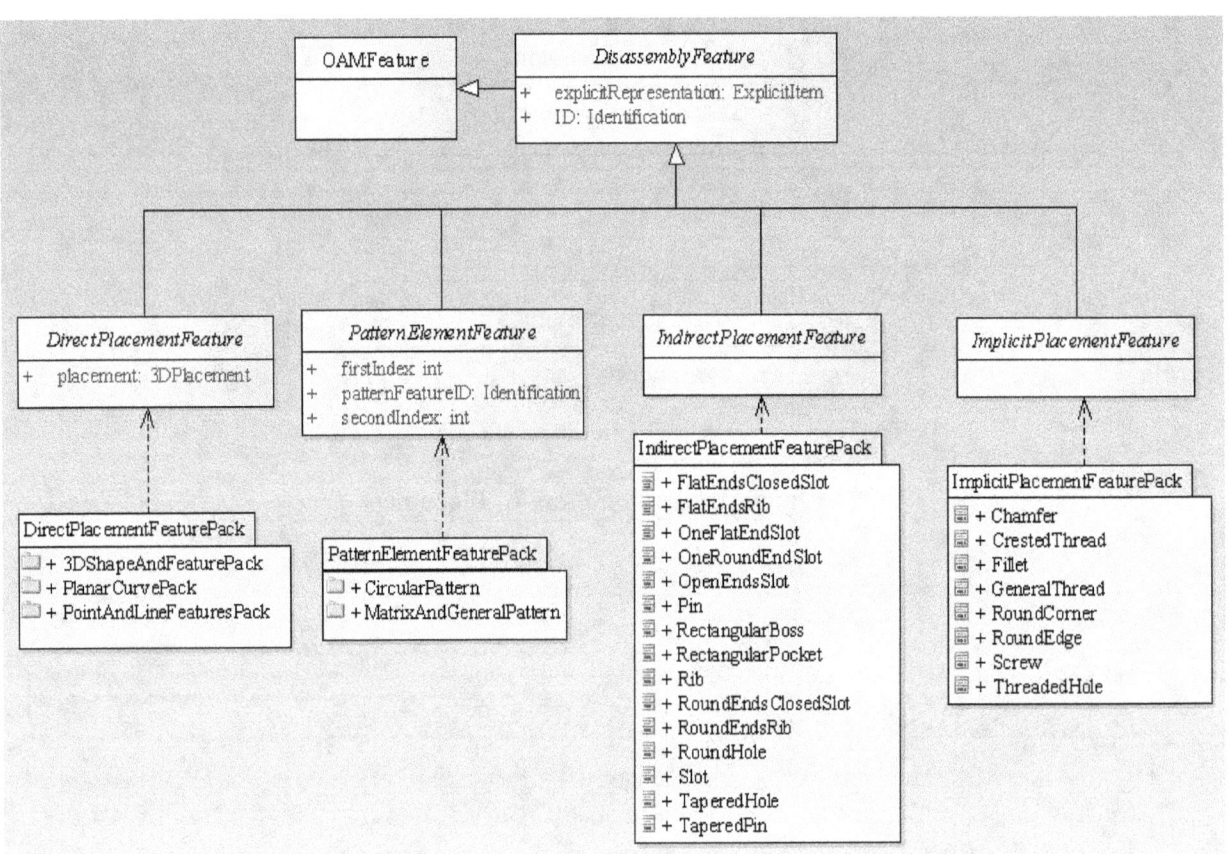

Figure 3. An Overview of Disassembly Feature Model
NOTE: class name in italic font in a UML class diagram means the class is abstract.

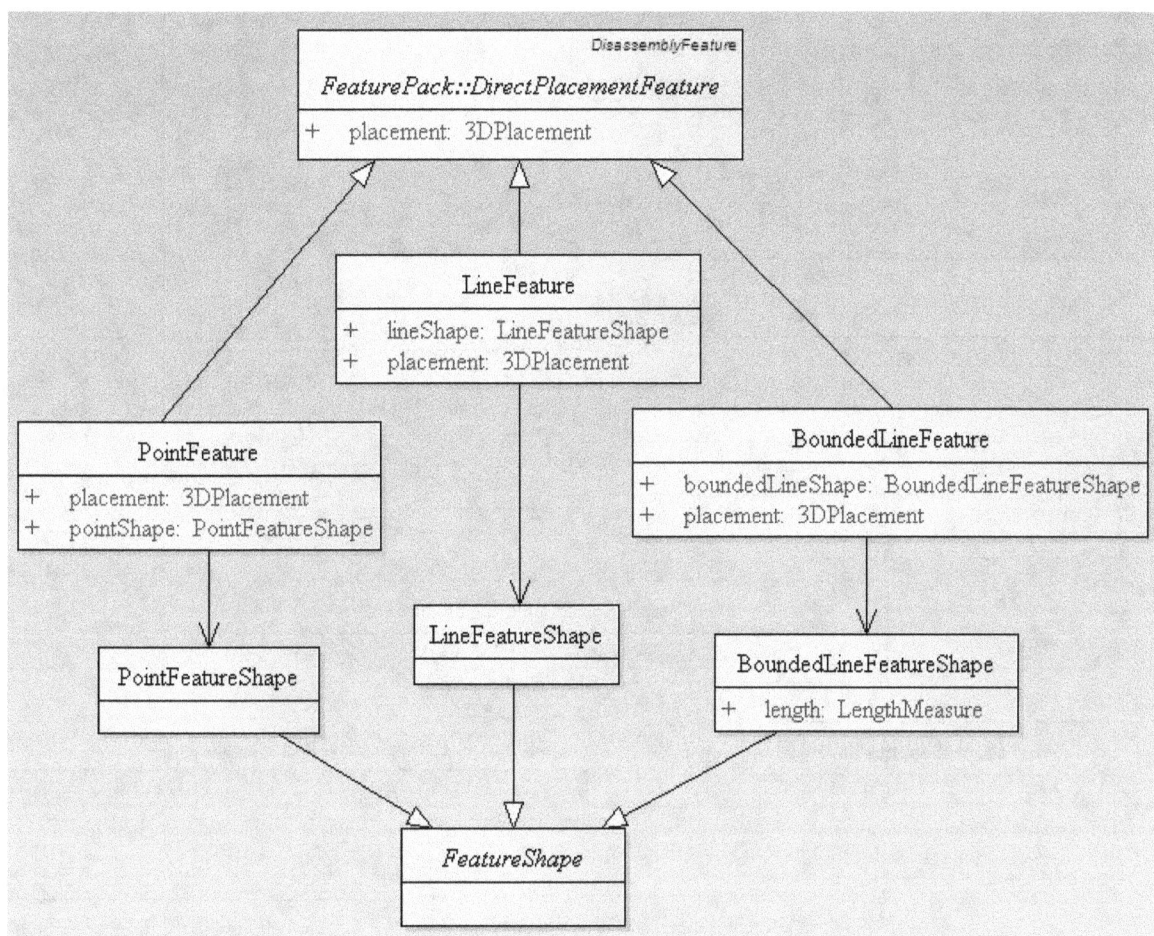

Figure 4. Class Diagram of Point and Line Features

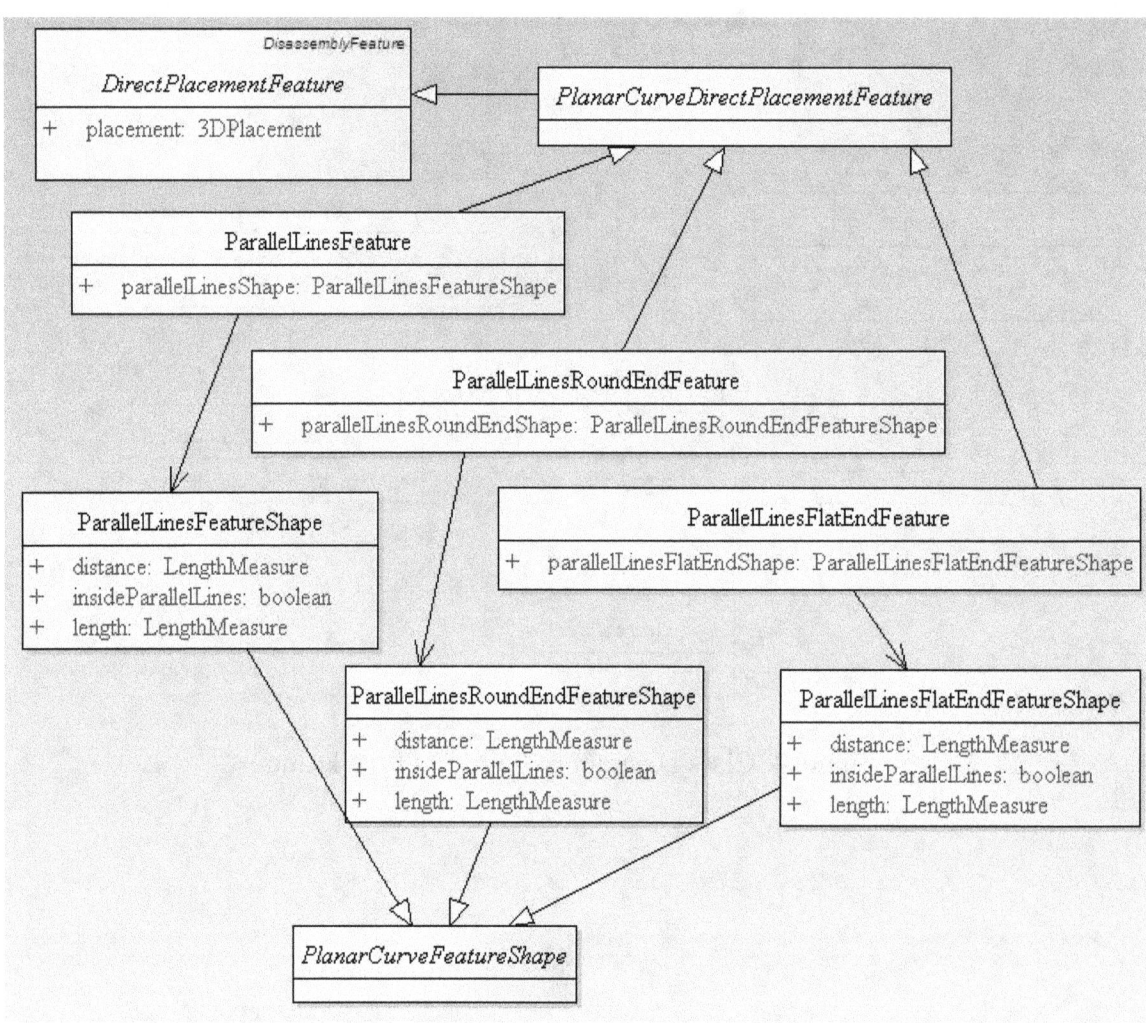

Figure 5. Class Diagram of Linear Features

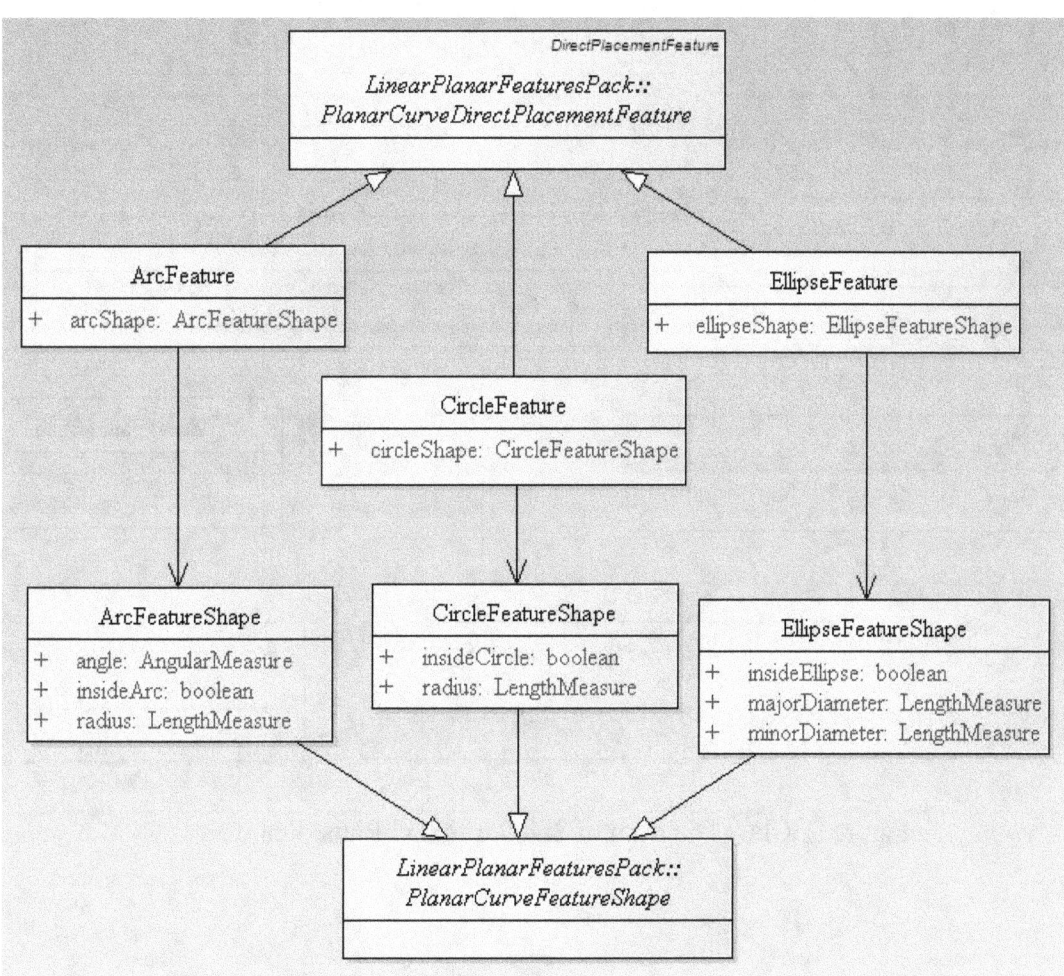

Figure 6. Class Diagram of Circular Features

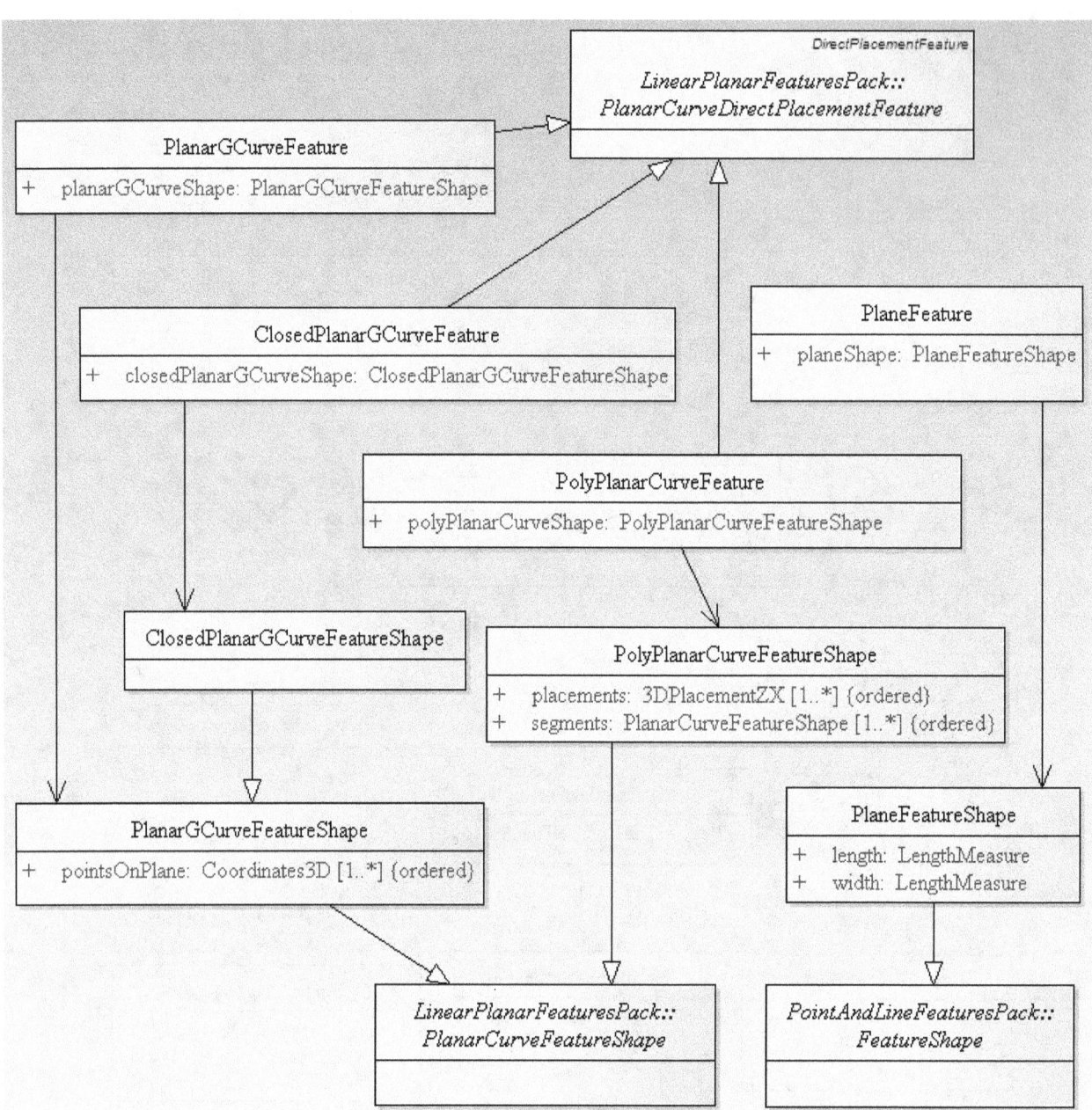

Figure 7. Class Diagram of 2D-Curve and Plane Features

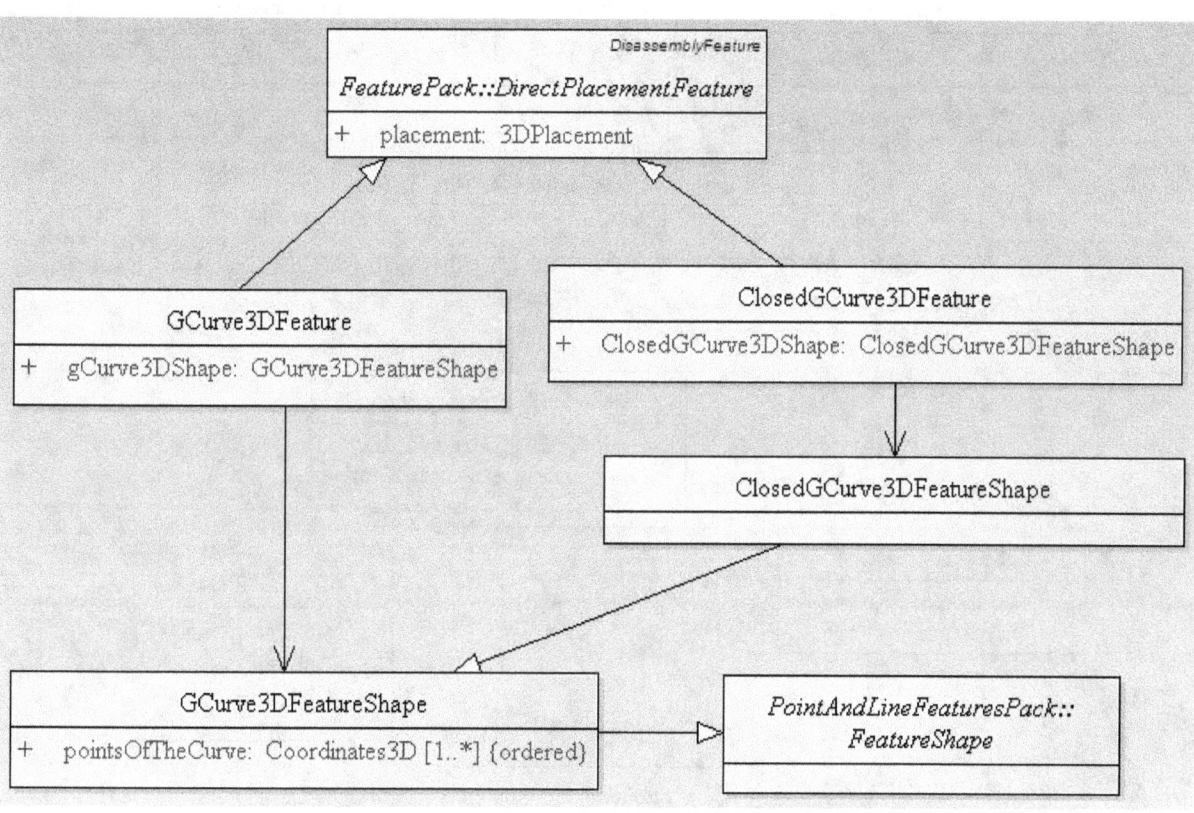

Figure 8. Class Diagram of 3D-General Curve Features

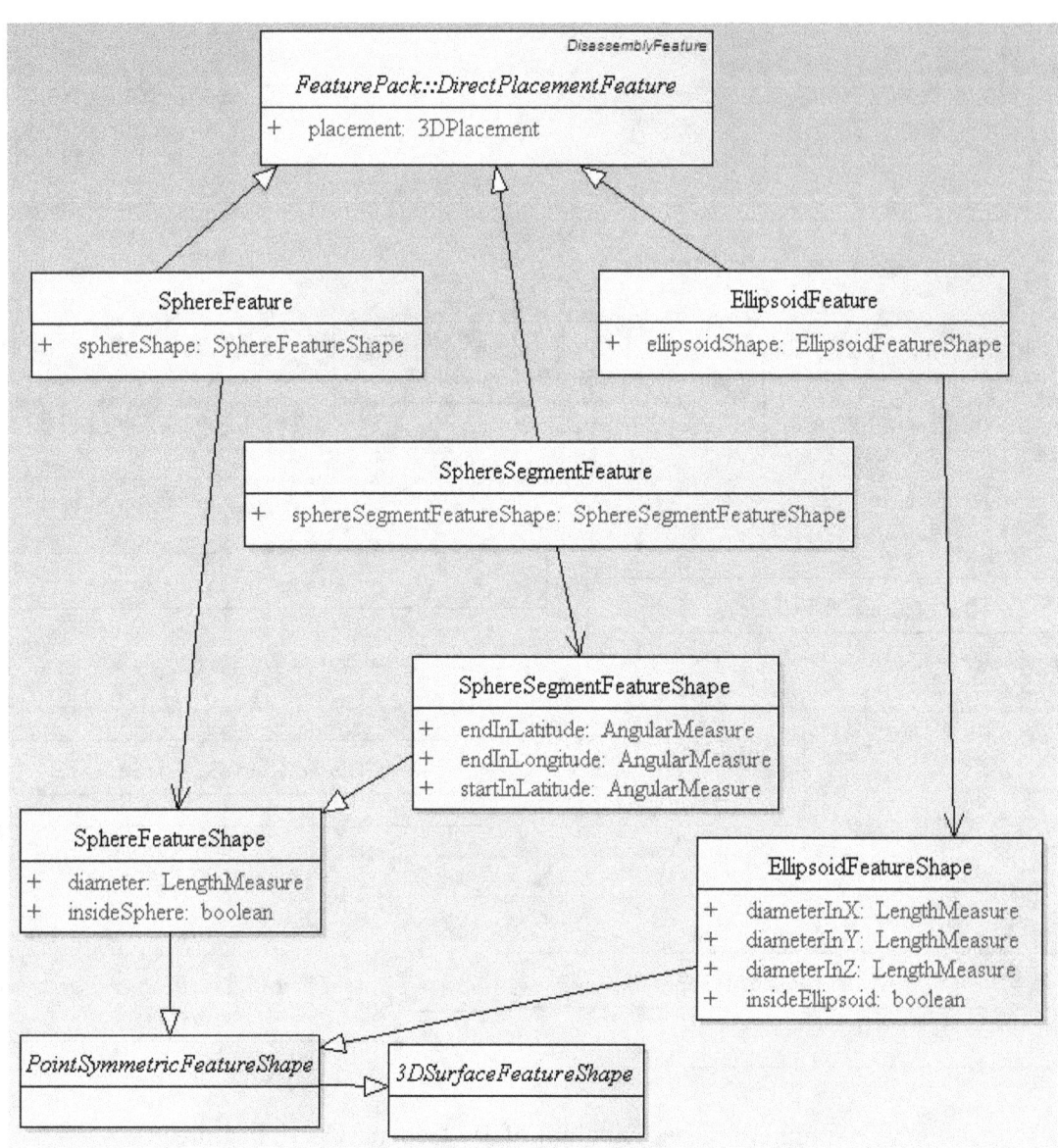

Figure 9. Class Diagram of Point Symmetric Features

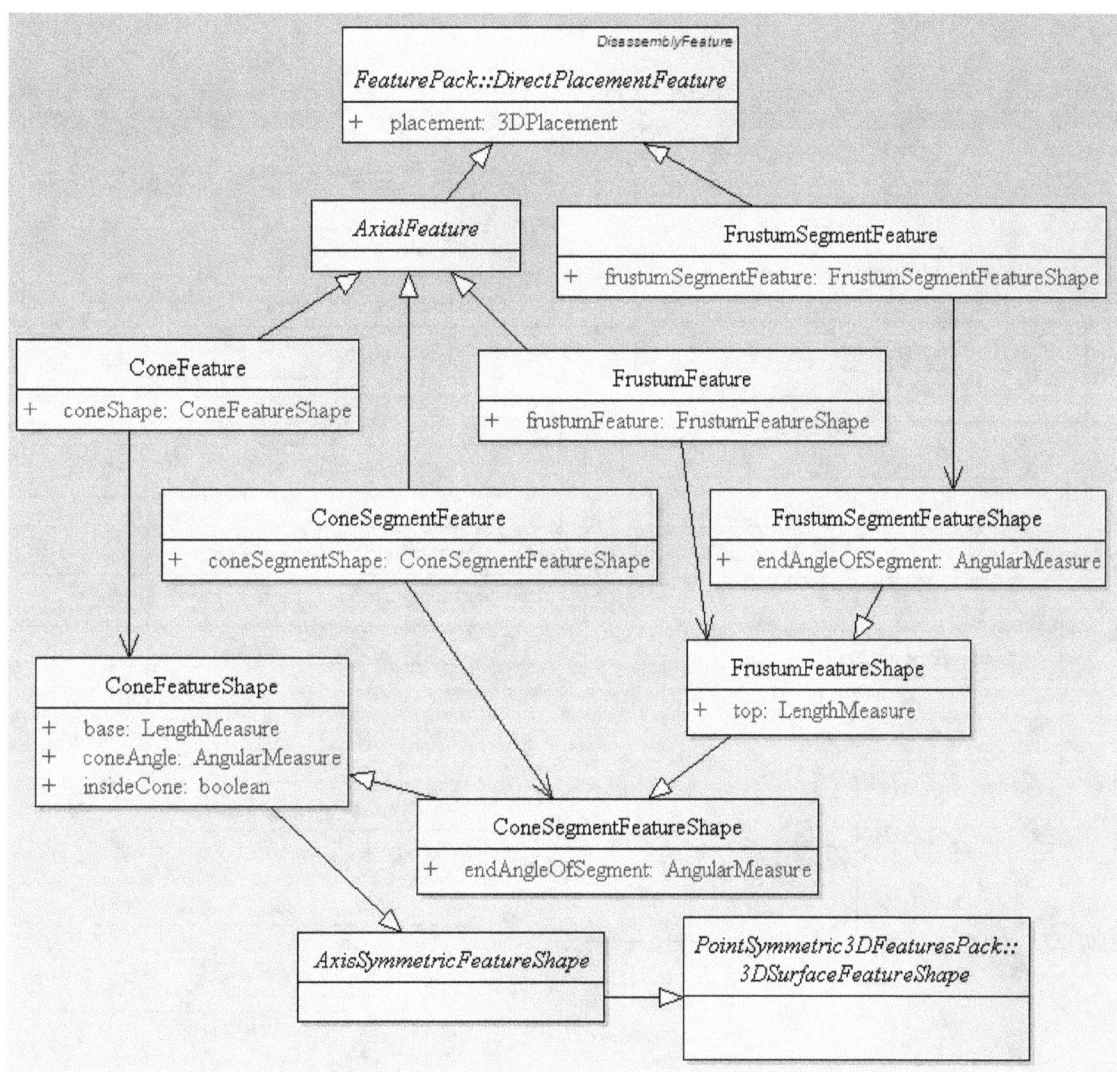

Figure 10. Class Diagram of Axial-Symmetric Features

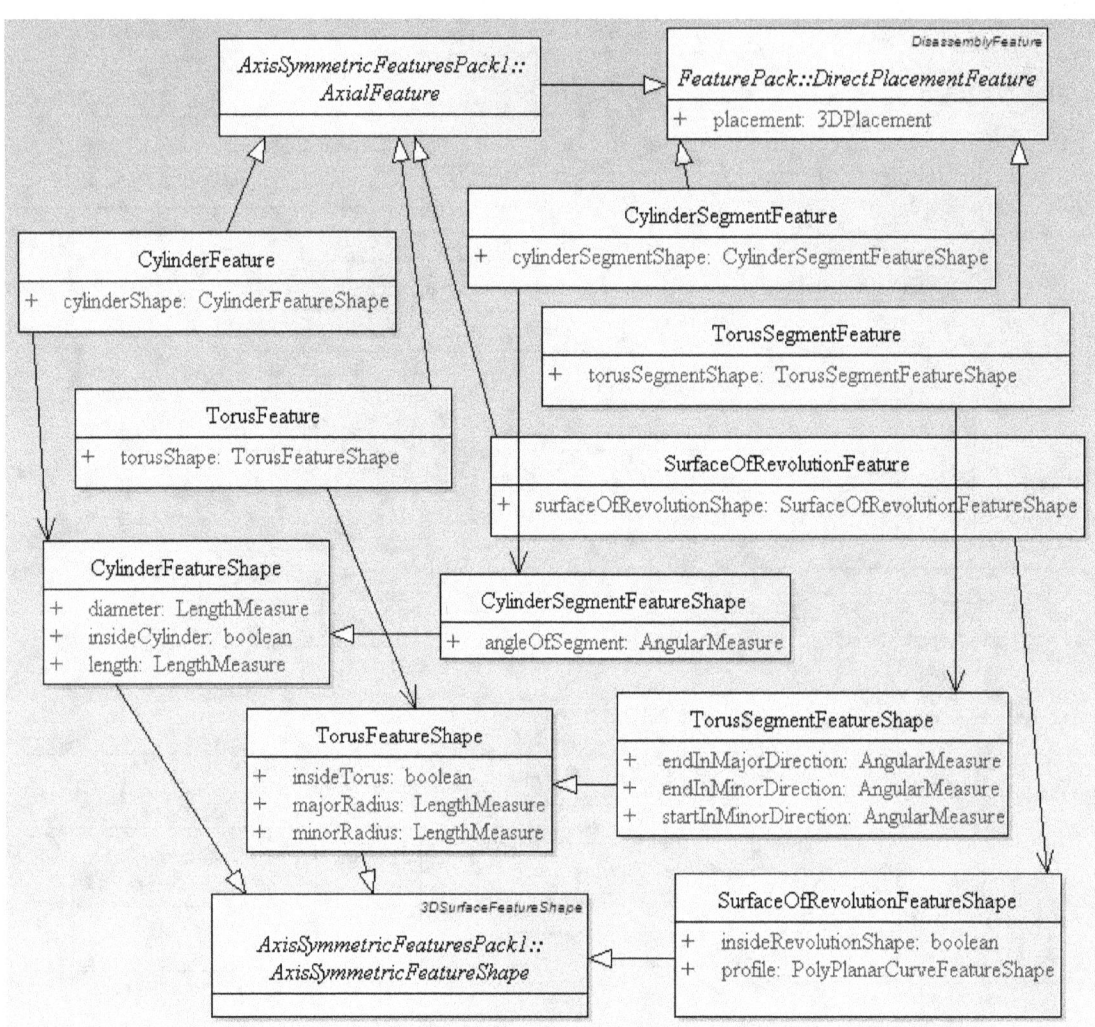

Figure 11. Class Diagram of Axial-Symmetric Features (cont.)

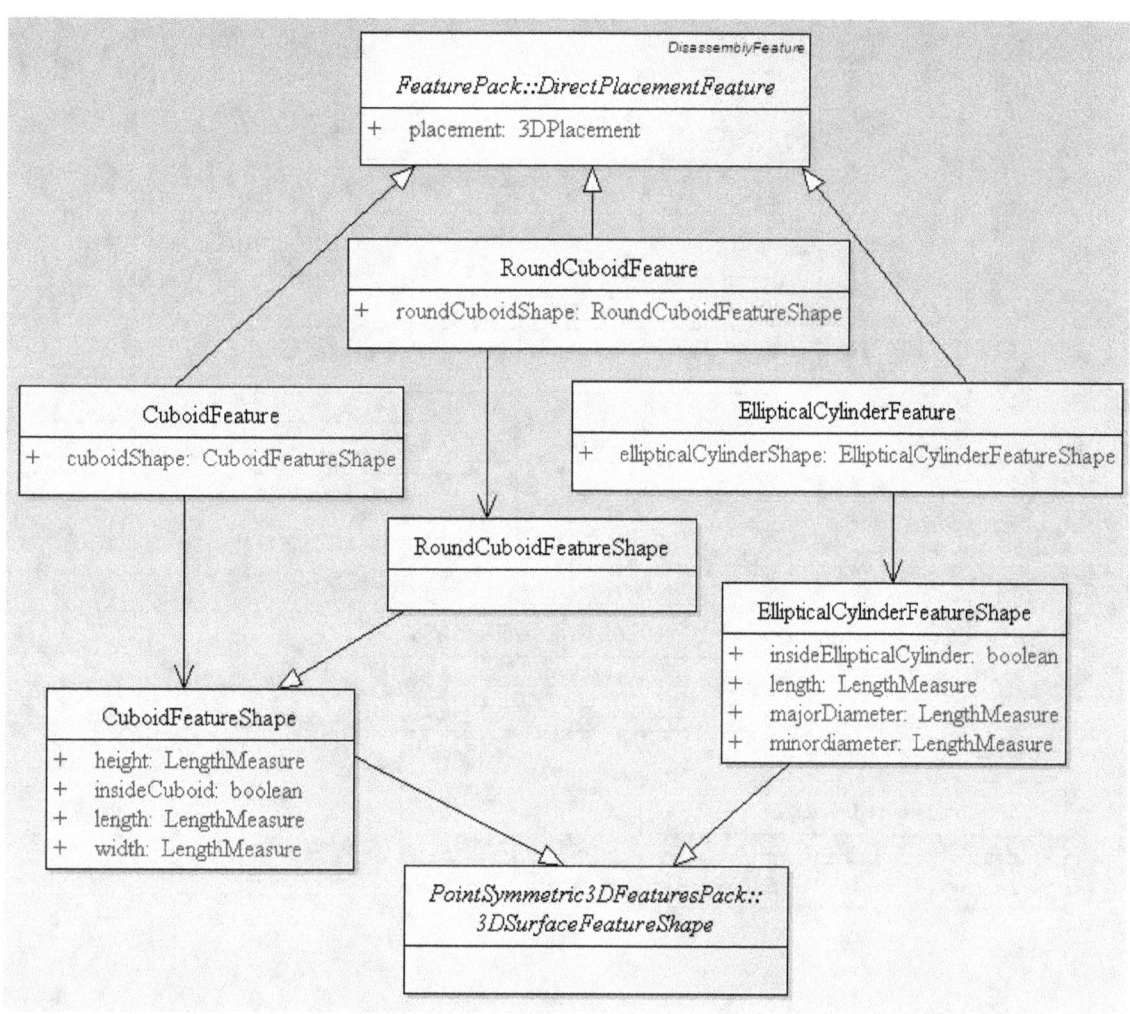

Figure 12. Class Diagram of Cuboid Features

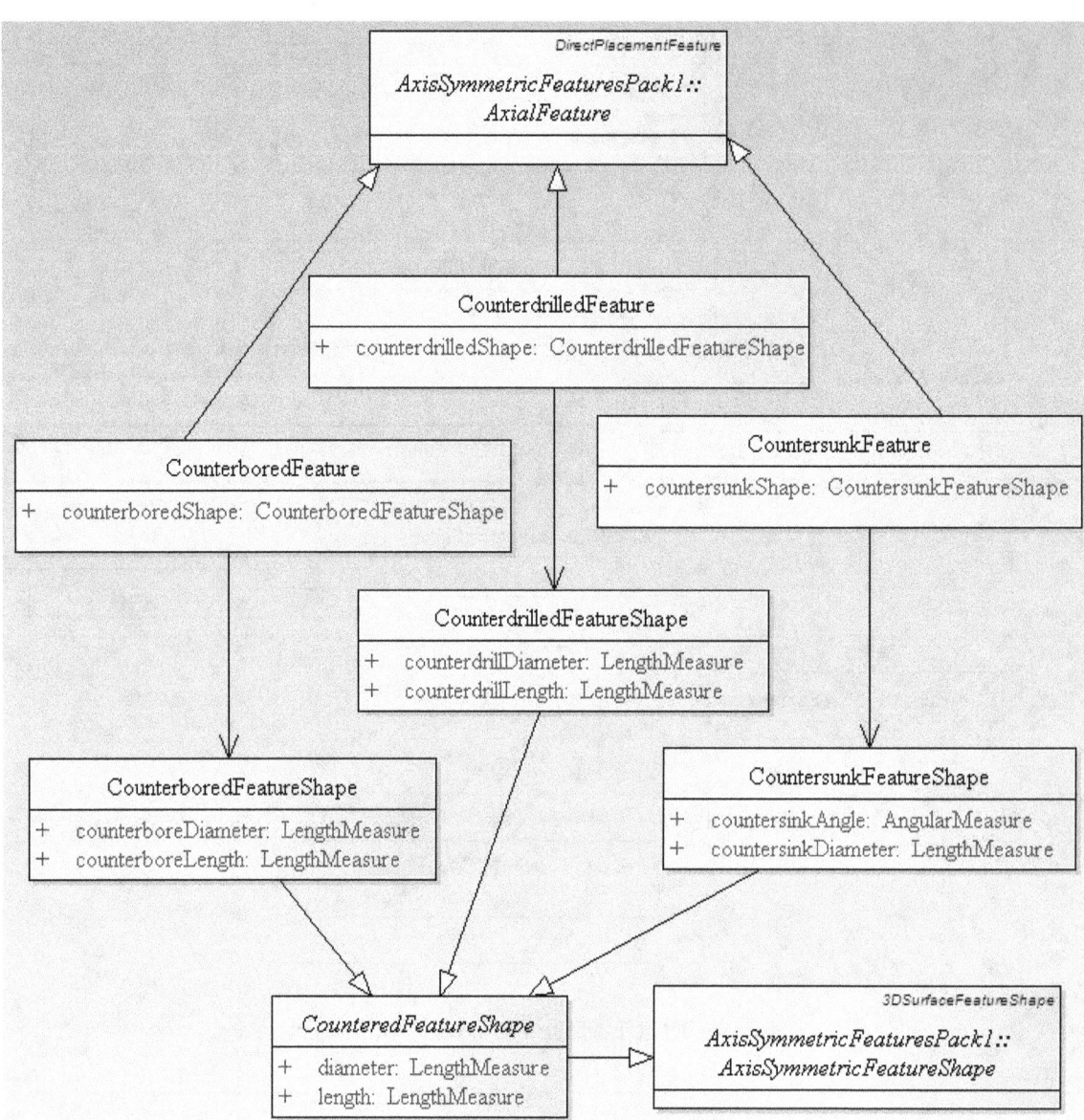

Figure 13. Class Diagram of Composite Features

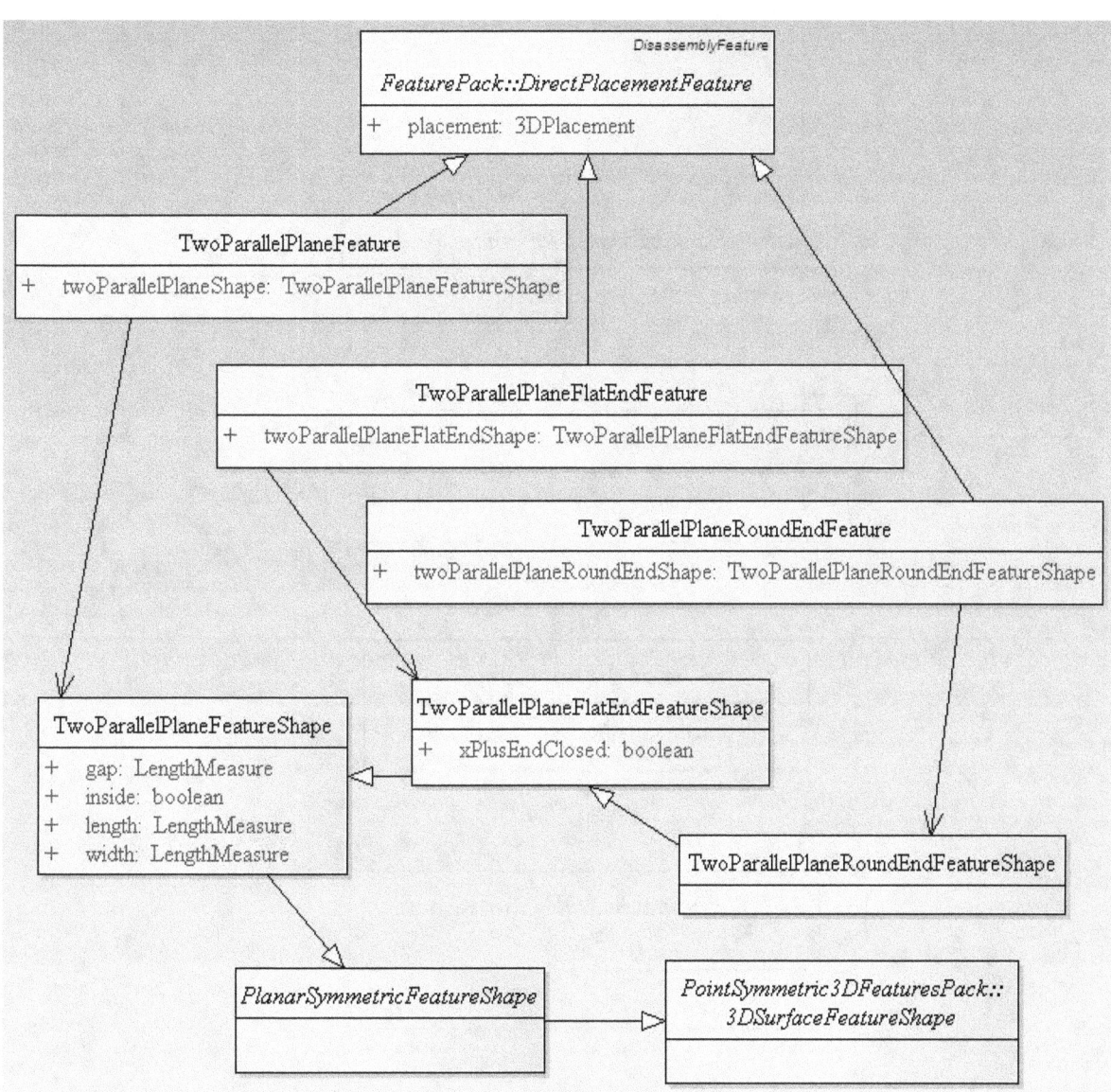

Figure 14. Class Diagram of Planar Features

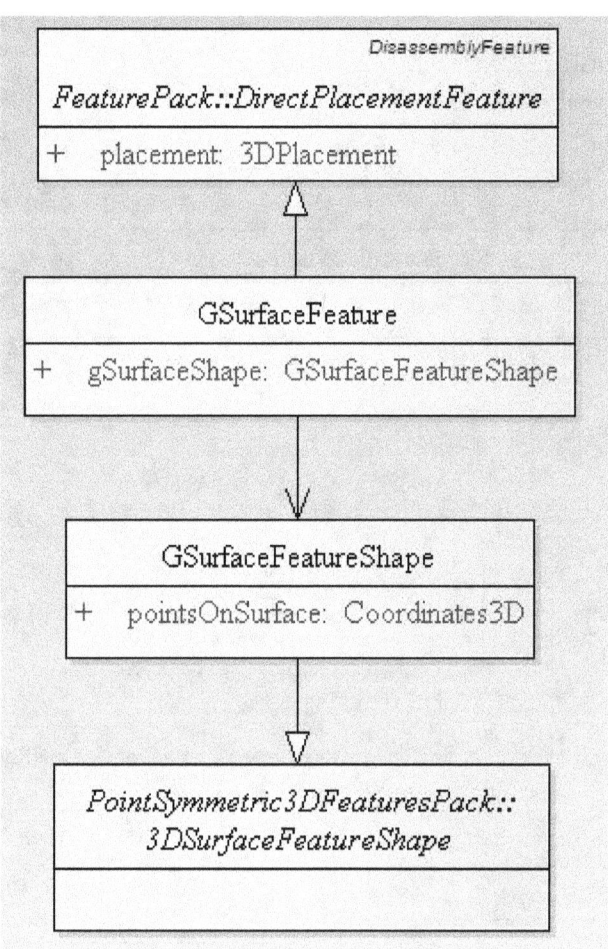

Figure 15. Class Diagram of 3D General Surface Features

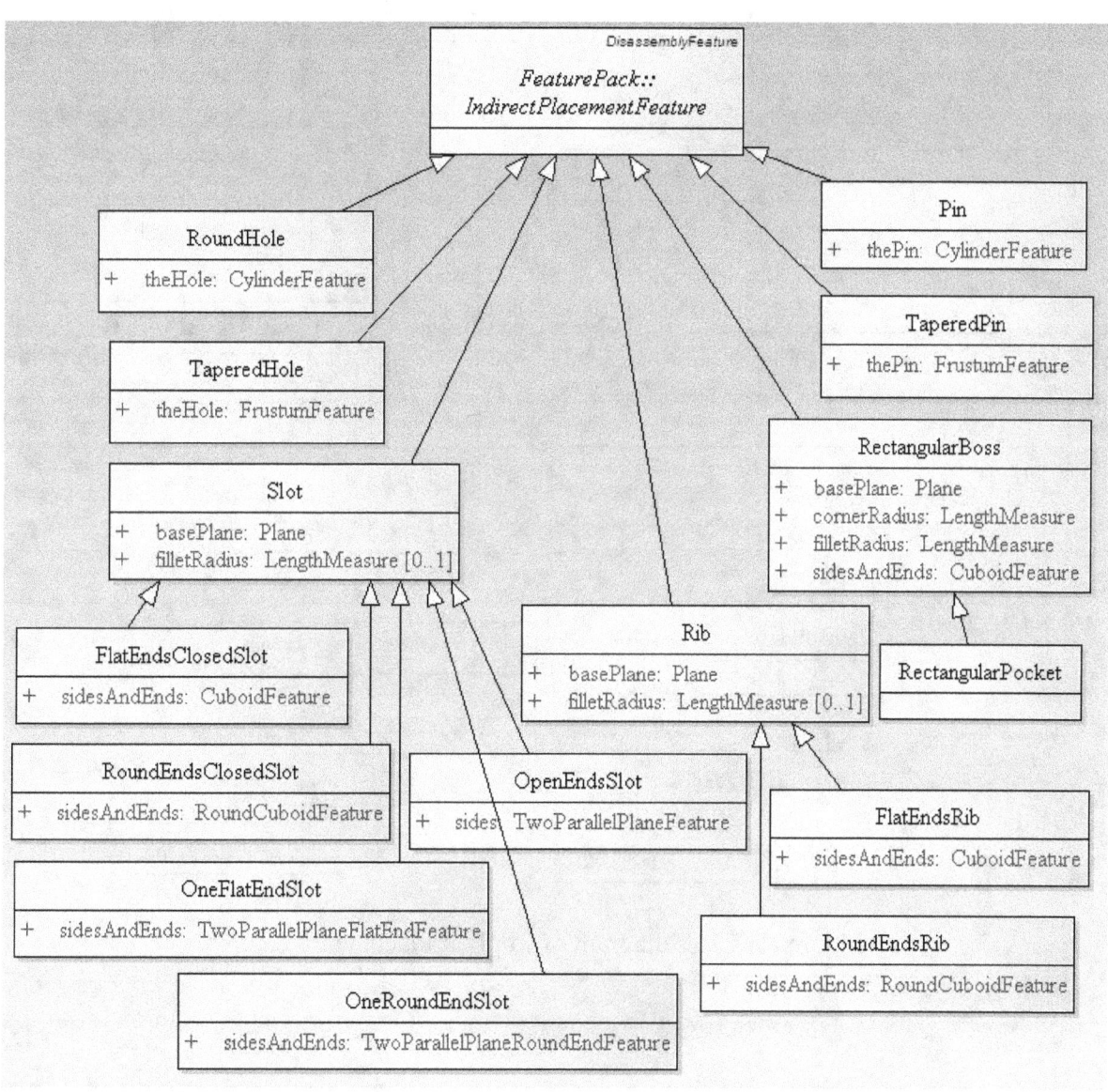

Figure 16. Class Diagram of Indirect Placement Features

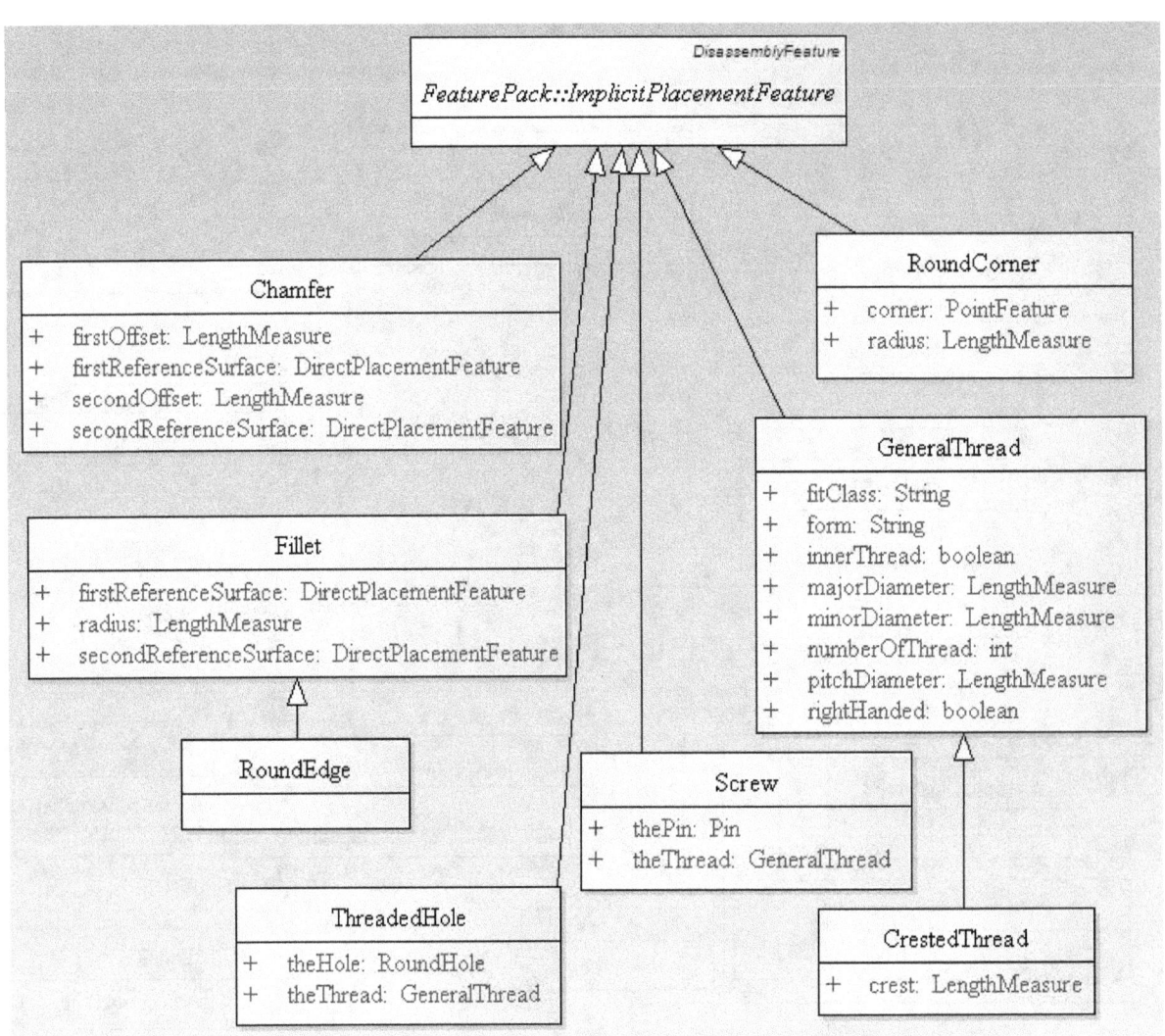

Figure 17. Class diagram of Implicit Placement Features

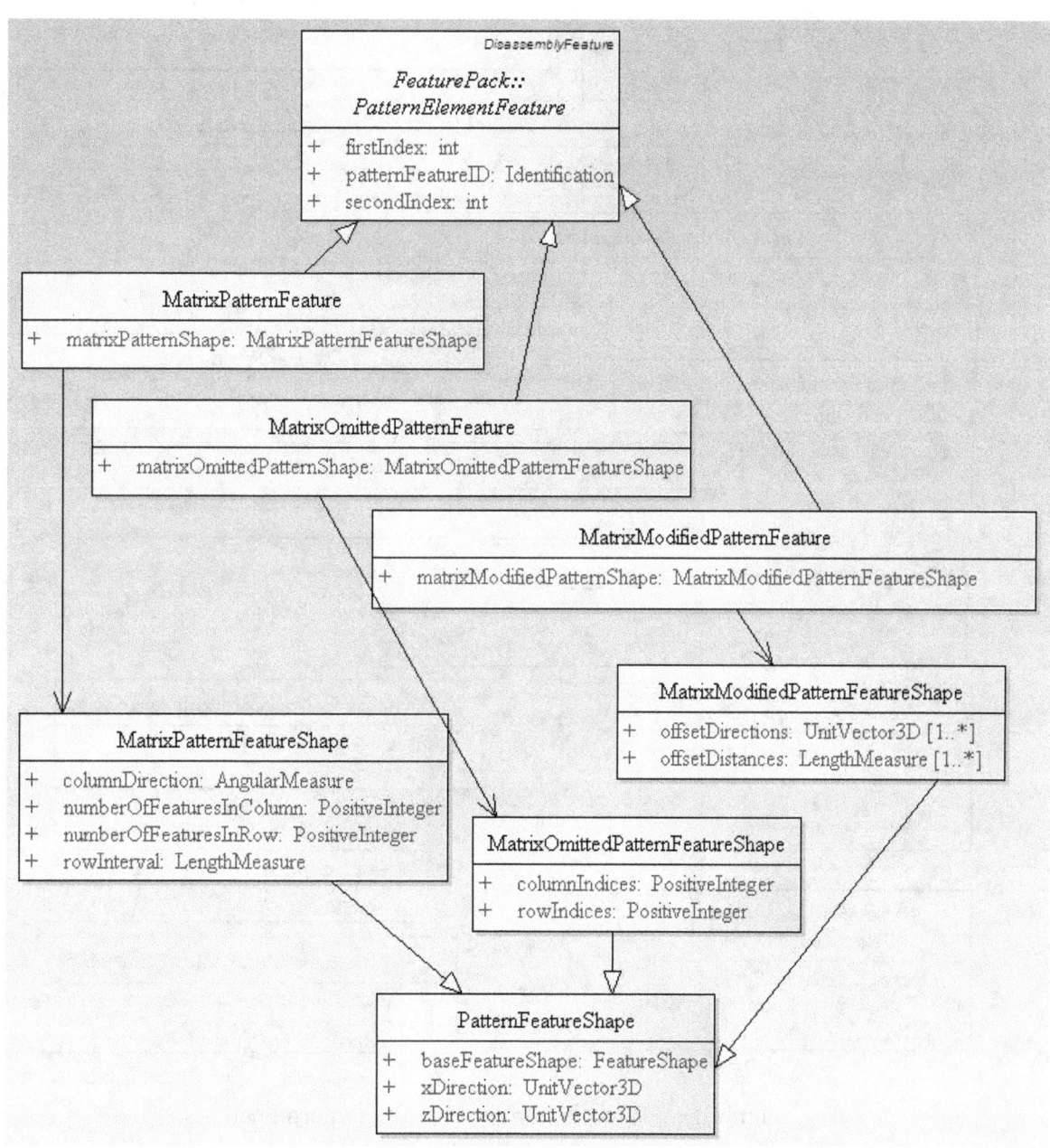

Figure 18. Class diagram of Matrix Pattern Features

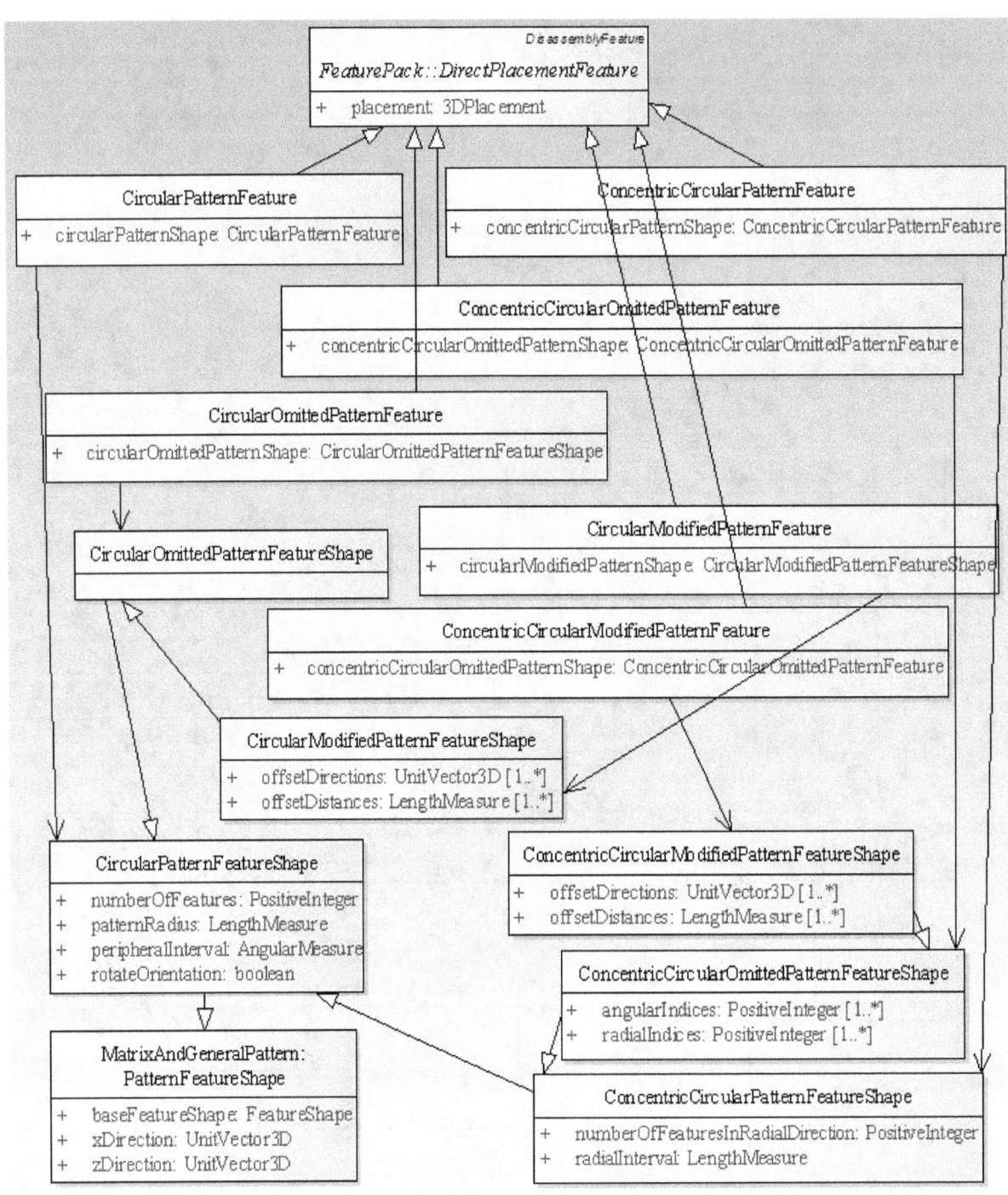

Figure 19. Class diagram of Circular Pattern Features